T0177716

Patterns in Mathematics Classroom Interaction

Patterns in Mathematics Classroom Interaction

A Conversation Analytic approach

JENNI INGRAM

OXFORD
UNIVERSITY PRESS

OXFORD
UNIVERSITY PRESS

Great Clarendon Street, Oxford, OX2 6DP,
United Kingdom

Oxford University Press is a department of the University of Oxford.
It furthers the University's objective of excellence in research, scholarship,
and education by publishing worldwide. Oxford is a registered trade mark of
Oxford University Press in the UK and in certain other countries

Published in the United States of America by Oxford University Press
198 Madison Avenue, New York, NY 10016, United States of America

British Library Cataloguing in Publication Data

Data available

Library of Congress Control Number: 2020952680

ISBN 978-0-19-886931-3

DOI: 10.1093/oso/9780198869313.001.0001

Printed and bound by
CPI Group (UK) Ltd, Croydon, CR0 4YY

Patterns in Mathematics Classroom Interaction

OXFORD
UNIVERSITY PRESS

Great Clarendon Street, Oxford, OX2 6DP,
United Kingdom

Oxford University Press is a department of the University of Oxford.
It furthers the University's objective of excellence in research, scholarship,
and education by publishing worldwide. Oxford is a registered trade mark of
Oxford University Press in the UK and in certain other countries

Published in the United States of America by Oxford University Press
198 Madison Avenue, New York, NY 10016, United States of America

British Library Cataloguing in Publication Data

Data available

Library of Congress Control Number: 2020952680

ISBN 978-0-19-886931-3

DOI: 10.1093/oso/9780198869313.001.0001

Printed and bound by
CPI Group (UK) Ltd, Croydon, CR0 4YY

Links to third party websites are provided by Oxford in good faith and
for information only. Oxford disclaims any responsibility for the materials
contained in any third party website referenced in this work.

Patterns in Mathematics Classroom Interaction

A Conversation Analytic approach

JENNI INGRAM

OXFORD
UNIVERSITY PRESS

Acknowledgements

This book reflects many of the conversations that have occurred at different times and places, in different classrooms, at research conferences, and with colleagues. I am indebted to all the teachers who opened up their classrooms and allowed me to video them teaching, or who offered up videos of their teaching. I am also grateful for the financial support from the John Fell Fund, which partly funded the collection of videos for one of the projects that the data and the analysis arose from. Many colleagues have made me stop and think; in particular, I would like to thank Nick Andrews, who helped me tease out what it was I wanted to say in my writing but has also done more to make me think about my practice as both a teacher educator and a researcher than anyone else. I would also like to thank Ann Childs, Katharine Burn, and Velda Elliott, who commented on drafts of my writing, prompted some of the insights I share in this book, and have continued to support me.

I would also like to thank my family, who gave me the space and time to write, and who inspired me to think more about mathematics teaching and learning. Through conversations with my husband, Jon, about mathematics and the teaching of mathematics, and the disagreements about how to teach quadratic equations to our son James during the pandemic, in which most of this book was written, I was able to see my data from different perspectives. Thanks also need to go to my children James, Daniel, and Lissie, who each gave me time to think and write, but also kept me grounded in the everyday. James, I am particularly glad that you were learning how to bake whilst I wrote this book.

Contents

List of Extracts

List of Figures

1

Introduction

Learning mathematics is a social and interactional endeavour and is the *raison d'être* of mathematics classrooms. Learning happens through interaction, not only between students and the mathematics, but necessarily between the teacher and the students, and between students. By looking at interactions we are looking not just for evidence of what students have learnt, but for the process of learning itself.

Learning mathematics is complex. It is not a linear or stable process that is the same for all. Learning happens over time, and we are continually building and developing what some would call our 'repertoires of meaning-making resources' (Hall 2018, 34). Some of these repertoires will be enduring across contexts, whether that is across our everyday lives and the mathematics classroom, across the different mathematics classrooms we experience in school, or across the different interactional contexts within a particular mathematics classroom environment. Our interactions, however, are context-specific and vary depending not only on who we are interacting with, but also on the practices and purposes that underlie our interaction. The learning of mathematics is intersubjectively negotiated and happens through the interactional structures and practices examined in this book.

A great deal of educational research focuses on intervening to identify 'what works', yet many of these interventions have been less successful when they have been 'scaled up' into ordinary classrooms. There is a great deal of mathematics learning taking place in classrooms, but teachers, curriculum designers, and school leaders continually seek to find ways in which students can learn more and learn better. Mathematics education research that focuses on classrooms has undergone enormous interdisciplinary growth over recent years. However, classroom interaction, whilst structured and patterned, is essentially unpredictable and rarely stable, which poses a challenge to researchers concerned with the processes of learning through interaction that are contingent upon the actions of teachers and students. Researchers who are interested in students' learning of mathematics come from a wide range of intellectual traditions and disciplinary roots, and the concepts, theories, and methodologies informing the research are drawn from fields as diverse as

Patterns in Mathematics Classroom Interaction: A Conversation Analytic approach. Jenni Ingram,
Oxford University Press (2021). © Jenni Ingram. DOI: 10.1093/oso/9780198869313.003.0001

anthropology, cognitive science, linguistics, sociology, and psychology, to name but a few.

This book takes a Conversation Analytic (CA) approach, with its roots in sociology but also drawing from linguistics and psychology, to focus on mathematics learning as it happens in the mathematics classroom. It also makes use of other perspectives, bringing these perspectives into dialogue with one another, integrating what is already known within mathematics education research about the process of learning mathematics to contribute more than the sum of the perspective-specific findings. CA is data-driven and analytically inductive, and findings using this approach are often descriptive, with claims substantiated through the sharing of transcripts that make visible the teachers' and students' actions. Learning mathematics can be considered at three interrelated levels of social activity. CA focuses on the micro level of interaction within classrooms but can also reveal the influences of the meso level of the sociocultural contexts of schools and classrooms and the macro level of ideological structures on the site of learning itself. With CA the analysis is grounded on how teachers and students themselves experience mathematics classroom interactions, particularly their choices and actions within these interactions (Lee 2010).

The classroom context is dynamic and complex, and is shaped by the teachers and students interacting within it. It is also goal and task oriented in that interactions and activities are planned for and designed to enable students to learn or develop. These goal orientations combine with interactional patterns and structures, creating specific interactional contexts (Seedhouse 2019). This mathematics classroom context can also be examined at different levels. Whilst most research into the learning and teaching of mathematics within the classroom focuses on the context of the classroom itself, there are also sub-varieties of interactional contexts within every classroom. It is these interactional contexts where pedagogy and interaction come together. The pedagogical focus influences the structure and organization of the interaction, and the structure of the interaction constrains and affords the pedagogical actions. The goal of a mathematics lesson is likely to be to learn some mathematics, but there are likely to also be other goals in play, such as developing students' social interaction skills, involving students in their own learning, or managing behaviour. The teacher and the students may or may not share these goals. These goals are achieved through teachers and students interacting with each other. Opportunities to learn mathematics vary between and within classrooms, and researching interactions enables us to better understand this process of learning. From a CA perspective, the nature of this interaction is key, both in terms of what is said and how it is said. By analysing these interactions in detail, we can see how teachers and students achieve different things through these interactions and consequently how the learning of mathematics is co-constructed.

Conversation analysis largely focuses on identifying structures and patterns within these interactions. These patterns are often unmarked and considered by many to be common sense, yet by paying attention to these structures and patterns we can see *how* students treat aspects of classroom interaction such as what it means to know, learn, or listen. By explicating these structures or rules as many researchers describe them, we can see how teachers can use them to be more effective. These structures are not prescriptive rules, they are instead interpretive resources which both teachers and students can make use of.

'Can you sit down please?' is (usually) interpreted by students as a request to sit down, not a question requiring a yes or no response. There are lots of choices that teachers make when choosing both what to say and how to say it, and these choices are made both consciously and unconsciously. Skilful teachers can use these choices to their advantage. Consider the following requests: 'open your book please', 'open your book thank you', and 'open your book'. The differences may seem purely semantic, or to relate solely to politeness, but saying 'thank you' usually follows the completion of a request and thus implies that the request will be granted. The 'please' also changes the imperative to a request. Conversation Analysis is interested in not only *what* is said in interaction but also *how* it is said and consequently what is being done with an interaction.

This book brings together a wide range of findings about mathematics classroom interaction that have resulted from studies using a CA approach. The research is based on three projects, all focused on mathematics classroom interaction over a period of ten years. Building on Mehan's (1979a) and later Cazden's (2001) detailed examination of common lesson structures and discourse patterns, I explore the structures and patterns that pervade mathematics classrooms and influence how students treat mathematics, the learning of mathematics, and the teaching of mathematics.

Conversation analysis focuses on talk-in-interaction, but this talk is not just considered to be about exchanging information; it is how we collaborate and mutually orient to each other in order to achieve meaningful communication (Hutchby & Wooffitt 1998). Teachers and students actively construct the mathematics classroom in which they interact. The analytic goal of CA research is to describe and make explicit the norms and procedures we use that help us to make sense of each other when we interact. It is interested in how we achieve intersubjectivity.

At times, the analysis can raise more questions than answers. Some events, situations, or cases are very rare, or particular to just one mathematics classroom. Yet illustrating these situations can help us to ask those questions, seek the right data, notice the implicit, which can support our understanding of the interactions between the teaching and learning of mathematics. The same can be true of deviant cases. One of the challenges that arises from using naturally occurring data is that you do not necessarily get sufficient examples to be able to notice a theme or pattern, let alone be able to make any kind of generalisation.

Where the Data Comes From

The data used in the analysis that has led to this book, and that is used to illustrate the different patterns and structures of mathematics classroom comes from projects that I have led over the past ten years. The projects all include data that come from video recordings of mathematics lessons from a wide range of schools and classrooms in England, with students aged somewhere between eleven and eighteen. This resulted in forty-two videos of mathematics lessons or parts of mathematics lessons. In half these videos I was present, sitting in the back of the classroom with a video camera. In the other half the teachers themselves recorded their lessons or asked a colleague to record the lesson. Some teachers recorded just one lesson; others recorded five or six lessons. The teachers chose which class to video, which lesson to video and how to teach the lesson, and to this extent the data is naturally occurring. BERA ethical guidelines for educational research (BERA 2004) were followed, with all teachers and students consenting to the video recordings. At the time of the video recordings I was not working with any of the teachers in my capacity as a teacher educator, though in some cases I worked with the teachers before or after the video recordings were made.

The way that these videos are shared with you, the reader, is through transcription. Conversation Analysis has a long tradition and specific approach to transcription that I describe in more detail in Chapter 2, but there is a balance between precision and detail, and readability. Transcription is an analytic process (Ingram & Elliott 2019) and the analysis of the videos involved transcribing and then watching and re-watching the videos alongside these transcripts. Whilst the examination of the patterns and structures I describe in this book resulted from detailed transcripts using the Jefferson transcription system (Hepburn & Bolden 2013) as illustrated in Extract 1-1, many of the details included in these transcriptions can make them difficult to read.

Since one of the main audiences for this book is researchers in mathematics education, perhaps with no experience of reading transcripts from CA research, I have in many places limited the transcription details included in the examples and illustrations to those details that are most relevant to the discussion at the time. So, for example, where pausing and hesitation are features of the analysis in focus, hesitation markers are included and pauses are measured to the nearest 0.1 s (using Audacity® version 2.2.2 and 2.3.2 (Audacity team 2018)). At other times, a simple verbatim transcript has been shared, as in Extract 1-2.

I have also used the convention of all teachers' names beginning with T and all student names beginning with S. Teacher names are consistent across extracts and uniquely refer to one of the teachers who kindly recorded their lesson(s). Student names are only consistent within an extract, and as gender is not a focus of the analysis (that is, the teachers and students are not making the gender of students relevant to the interaction) in most cases the gender of the student pseudonym may not necessarily match the gender of the student in the video. There are two exceptions to this, where gender is oriented to by

```
54  Tyler:    yea:h? (.) but ↑what fractio::n (0.8) what
                >fraction of that< triangle have I shaded
                (0.8) what fraction of that triangle have I
                actually sha::ded (.) Simon,
55  Simon:    um (.) w- ↑half?
56            (0.6)
57  Tyler:    >have I shaded a ↑ha::lf
```

Extract 1-1 Jefferson transcript of classroom interaction.

```
54  Tyler:    yeah? (.) but what fraction (0.8) what
                fraction of that triangle have I shaded (0.8)
                what fraction of that triangle have I
                actually shaded (.) Simon,
55  Simon:    um (.) half?
56            (0.6)
57  Tyler:    have I shaded a half
```

Extract 1-2 Verbatim transcript of interaction from Extract 1-1.

the teacher and the student, and in these two situations the gender of the student has been preserved.

Structure of the Book

Chapter 2 outlines the origins, principles, and tools of Conversation Analysis, as well as summarizing some of the key structures and patterns on interaction that are the building blocks of any CA analysis. There are several books available that outline CA in more depth for the interested reader, and I have been necessarily brief, focusing only on those aspects that are drawn upon in the rest of the book. More thorough recent introductions can be found in Sidnell (2010) or ten Have (2007), and for more detail and examples of CA in a range of disciplines and fields the Handbook of Conversation Analysis (Sidnell & Stivers 2012) provides a comprehensive overview.

The remainder of the book is split into two main parts. In the first part I focus on structural regularities and patterns, such as turn-taking and repair, that are used by CA researchers to examine interaction in a range of contexts, including classrooms (Gardner 2019; Macbeth 2004; McHoul 1978, 1990). Chapter 3 focuses on turn-taking and Chapter 4 focuses on repair and trouble in interactions. In each of these chapters I begin by examining the patterns of turn-taking and repair within mathematics classroom interaction, highlighting structures that are consistent across interactional contexts, as well as exploring variations and the influence these have on the learning and teaching of mathematics. At the start of each chapter I have tried to stay faithful to the CA descriptive principles by describing the interactional and learning process as they are oriented to by mathematics teachers and students. Yet description can only take us so far. Towards the end of these chapters I also draw on research within mathematics education to consider how these structures influence learning, whilst still treating learning as a social accomplishment that takes place over time.

In Chapters 5 and 6, the focus shifts to the processes associated with learning mathematics and draws upon the tools of CA in order to contribute a more nuanced understanding of what it means to learn mathematics. The emphasis continues to be on how teachers and students themselves treat these processes as they interact in the mathematics classroom, but with a focus on the mathematical tasks, activities, and behaviours that students and teachers engage in. However, this in itself deviates from ethnomethodological principles, as teachers and students often do not make distinctions between activities or actions being mathematical or not. The distinction between what is a mathematical explanation and what is an explanation is often not made in classroom interactions. My own experiences as a mathematics teacher, mathematics education researcher, and teacher educator are often drawn upon in the analyses in these two chapters.

Throughout I have treated learning mathematics as a way of acting: mathematics is something that you do, not just something you know.

2

Conversation Analysis

The analyses of mathematics classroom interaction explored in the following chapters all use Conversation Analysis (CA). In this chapter I will first describe the context in which CA developed as an approach to analysing interaction, which leads to some key basic principles that underpin any research using CA. Conversation Analysis is both a theoretical perspective on interactions and a method for researching these interactions. As a theoretical perspective it views each turn in interactions as social actions: it looks at what we are doing with what we say. In contrast to other popular approaches to studying interaction, it is not about what people mean in terms of what is going on inside of their head, but what they mean through what they do. Conversation Analysis focuses on the unfolding of the interaction. This includes *what* is said or done, but also *how* it is said or done. The questions driving the analysis are 'Why that now?' (Bilmes 1985; Schegloff 2007): why did a particular person say that, in this particular way, and at this point in the interaction. Furthermore, what is being done by how things are said, and how are the other people in the interaction treating what has been said? At the core of CA is a concern with how participants in interaction engage in intersubjectivity through the interactional work they do in producing their own turns at talk, which also makes visible their understanding of the turns of others (Heritage 1984). It is not a linguistic or cognitive analysis of what is being said, though CA can be combined with linguistic and sociological approaches, as I illustrate in Chapters 5 and 6.

This focus on how things are said leads to a focus on the structure or interactional architecture (Seedhouse 2004) of interactions. There are some well-established interactional structures of ordinary conversation that are also relevant to classroom interaction, and I will illustrate these below with data from mathematics classrooms. Chapter 3 focuses specifically on the structure of turn-taking, how different turn-taking structures exist within classrooms and the effects these have on mathematics classroom interactions. Conversation Analysis in particular distinguishes between ordinary conversation and institutional discourse, and these interactional structures demonstrate some fundamental differences in structure between ordinary

Patterns in Mathematics Classroom Interaction: A Conversation Analytic approach. Jenni Ingram,
Oxford University Press (2021). © Jenni Ingram. DOI: 10.1093/oso/9780198869313.003.0002

conversation and mathematics classroom interaction. The differences between ordinary conversation and classroom interaction reveal the underlying purpose of classroom interactions: learning. These differences include how turns at talk are allocated, how repair is conducted, how turns are constructed and sequentially organized, as well as the structure of instructions and explanations.

These interactional structures are often implicit and unnoticed by the people interacting, but they are what we use in our interactions to help us make sense of what is going on. It is also these structures that make the institutional contexts in which we are interacting, such as in classrooms, relevant to our analysis. Whilst this may seem a little odd, how context is treated within CA is markedly different from other discursive approaches, as I discuss below. In particular, we cannot claim that the fact that the interaction takes place in a classroom is relevant to our analysis unless the participants (usually teachers and students) show that it is relevant. One way in which they do this is by using the interactional structures found in classrooms that differ from other contexts, such as the structure of classroom turn-taking.

I often use the word 'structure' instead of 'rules'. This is because, for many, the word 'rule' conjures up feelings of constraint and restriction. Rules are there to be followed. Some school rules are intended to help activities in school run smoothly, and others, such as those dealing with uniform, may be more about developing a sense of community and a school identity; they are expected to be held in respect and venerated, not broken. Those who break the rules can expect to be sanctioned. In a similar way, if you break the rules of turn-taking described below, you can also expect to be sanctioned. These rules also enable classroom interactions to run smoothly and accomplish a range of actions. Yet these turn-taking rules do not only constrain what can be done. They can also be resources that teachers and students can use. Whilst they act to constrain liberties, they serve as liberating constraints. There is no rulebook to be learnt. Instead these rules are often implicit and acquired over time through the interactions that they govern. In this respect, I prefer the terms 'structures' or 'norms'. These terms convey the idea of patterns and routines that might be labelled as rules, but instead of breaking them, we deviate, adapt, and reconstruct them. Yet these terms can also be problematic. The word 'norms' is often used at a higher or macro level where we are referring to social or cultural norms, and has been used widely in the mathematics education literature, particularly since Yackel's and Cobb's (1996) demarcation of sociomathematical norms, but often without definition or consistency.

Structure describes an organization, configuration, and arrangement of different elements. In mathematics, structure is about relationships between mathematical objects, parts, or elements (Dörfler 2016). In the case of turn-taking, it is the organization of who can take turns and what these turns can contain that is structured. Mason talks about structure having an 'architectural quality' (in Venkat et al. 2019), yet the metaphorical use of architectural here, or the idea of 'interactional architecture' as described by Seedhouse (2004), implies some kind of conscious design. Yet the rules described below and in later chapters have evolved over time. They have been established, negotiated, and re-negotiated, both consciously and deliberately and unconsciously without intention. Whilst these rules or norms are fluid, adaptable, and dynamic, there is also an underlying structure that enables classroom interactions to flow unhindered. It is this structure that is of most interest to CA researchers.

Conversation Analysis is often treated by many researchers as a set of methods for data analysis, but it is far more than this. It is embedded within a strong theoretical background and has a distinctive methodology. There are key principles that underpin all true CA research which drive the decisions around the research questions that can be addressed, what data will be collected, and how data will be collected, as well as how any analysis is conducted and reported. It is not just the analysis of conversation or a type of discourse analysis. Whilst there are particular structures that are often of interest to CA researchers, such as turn-taking and repair, there are also principles which guide study design, data collection, and the presentation of findings, which leads to many researchers using the term EMCA (ethnomethodological Conversation Analysis) to emphasize that their theoretical and methodological approach is underpinned by these principles. In the rest of this chapter I will outline these principles and their effect on research methodology. I will also outline the interactional structures widely used in CA research that are particularly relevant to the analysis of classroom interactions. Each of these are outlined in this chapter with illustrations from my own data collected from a range of mathematics classrooms. There are several interactional structures that I do not describe as they are not drawn upon in the analysis presented in the later chapters, but the interested reader will find many others in Schegloff's primer (2007) and in *The Handbook of Conversation Analysis* (Sidnell & Stivers 2012).

Origins of Conversation Analysis

Conversation Analysis was developed by Harvey Sacks and his colleagues Emanuel Schegloff and Gail Jefferson at the University of California. Sacks was heavily influenced by the work of Harold Garfinkel (1967), a leading figure in ethnomethodology, and the sociologist Erving Goffman (1981), who studied symbolic interactionism. Goffman and Garfinkel developed a new approach to studying interaction by studying how participants themselves make sense of each other in interaction itself. Sacks, Schegloff, and Jefferson then developed a method for systematically studying interaction using this approach (Enfield & Sidnell 2014). The influence of Goffman and Garfinkel is evident in the principles that guide CA research and its approach to data analysis. The development of CA also coincided with developments in technology which made audio recordings of interactional data far easier to collect. Garfinkel promoted the idea of collecting and analysing naturally occurring data in contrast to the prevalent use of laboratory settings and experiments in both sociology and the study of language at that time. This idea of collecting naturally occurring data continues to be a key feature of CA research, though the meaning of 'naturally occurring' varies depending upon the interactional contexts being researched. Central to CA and ethnomethodology is this focus on naturally occurring data, but also a 'naturalistic' approach to this data (Macbeth 2003). This is similar to the idea of 'accounts of' before 'accounting for' within mathematics education (e.g. Coles 2013; Mason 2002, 2012), where the researcher (or teacher) refrains from making any inference or judgement of what is happening within an interaction and thus imposing external categories or values upon the interaction. Instead the analysis examines those categories or values that the participants themselves are orienting to in the interaction itself.

Conversation Analysis is a 'naturalistic observational discipline that could deal with the details of social action rigorously, empirically and formally' (Schegloff & Sacks 1973, 289–90). Sacks, building on the work of Garfinkel, argued that interaction is systematically organized and ordered, and challenged the idea that ordinary conversation was too chaotic to be studied. He investigated normative patterns of interaction and how participants design their turns for others in the interaction, making use of these patterns or structures. Whilst there are differences in these normative patterns depending upon the context within which they occur, i.e. classroom talk is different to conversations around the dinner table, these normative patterns and how people orient to them often tells us a great deal about the interactional context in which they occur.

A more detailed history of the development of CA can be found in Heritage (1984). Here, I emphasize those aspects that influence how researchers use CA to analyse mathematics classroom interactions. Mathematics classrooms are an institutional context, and whilst Sack's early research focused on the analysis of calls to a suicide helpline, the development of many of the interactional structures described below arose from the study of ordinary conversations. Whilst there is some debate around the use of CA in institutional contexts, particularly in the early stages of its use as a research approach, recently CA is being used in a wide variety of institutional contexts; this book builds on and makes use of this more recent research. Readers interested in these debates are invited to read Paul Drew and John Heritage's book *Talk at Work* (1992).

Ethnomethodology

Ethnomethodology is a sociological approach which studies the principles on which we base our social actions. It is the study of 'the body of common-sense knowledge and the range of procedures and considerations by means of which the ordinary members of society make sense of, find their way about in, and act on the circumstances in which they find themselves' (Heritage 1984, 4). When we interact, we select only certain pieces of information and we try to organize this information into some sort of underlying pattern that enables both us and other participants to make sense of the interaction. Within this, CA focuses more narrowly on studying interactions between people rather than in texts. Originally it exclusively concentrated on interaction through language, but more recently the analysis has included gestures and body positioning, including eye gaze within the interaction.

Garfinkel argued that we use normative principles when we interact with the world that enable us to both display our actions and allow others to make sense of them. These principles are seen-but-unnoticed, and one role of the researcher is to uncover and describe these principles. Garfinkel used breaching experiments to uncover these principles in his own research. However, in his attempts to breach the norms of interaction he found that people adjusted their interactions to try to make sense of the interaction as if everyone was following the same norms, but they also felt upset and hostile towards the people they were interacting with because of their failure to cooperate, which 'threatens the very possibility of mutual understanding' (Heritage 1984, 43). Today these norms of interaction are studied by examining how people react when these norms are deviated from naturally in conversations, rather than through an experimental design. This is often referred to as deviant case analysis; these cases are of particular interest, as it is when things deviate from the norms that the norm itself becomes explicit. Deviant cases 'often serve to demonstrate the normativity of practices' (Heritage 1995 cited in Seedhouse 2019).

One key principle of all ethnomethodological approaches is the focus on the perspective of the participants, rather than the perspective of the researcher. What is relevant to any analysis is what is made relevant by the participants in how they interact. As Macbeth (2003, 241) states, the participants themselves are 'the first analysts on the scene'. Interactions contain 'everything relevant for analysis' (Cameron 2001, 88), and we do not necessarily draw upon other contextual aspects such as the gender or status of the participants unless these are evident in the interaction itself. There is a commitment to investigating how, through interaction, we establish, sustain, and change what it means to participate (Sahlström 2009).

In the study of classroom interaction, we need to consider both the teacher and the students as it is concerned with how we coordinate action and meaning (Abrahamson, Flood, Miele, & Siu 2019) visibly in interaction. This is a key difference between ethnomethodological approaches and others in educational research that depend upon researchers' codes or participants' self-reports. In the study of classroom interaction we need to consider both the teacher and the students, as interactions are co-constructed by both. This focus on the perspective of participants means that coding data from a developed framework is avoided, as this displays the researcher's interpretation of the interaction rather than the participants'. This interpretation or understanding of the interaction is often revealed through participants' actions, including what they say and how they say it. So, for example, teachers often give instructions, ask questions, and evaluate or assess when they interact in classrooms, whereas students generally answer questions and follow instructions. When a teacher asks a question and this is followed by a student answering the question without any complaints or resistance, both the teacher and student are showing that it is normative for a teacher to ask a question and for a student to answer a question. There is no challenging of these actions. It is the observation of this in classroom interactions that tells us as researchers that teachers and students have these roles as question askers and question answerers, not our pre-determined categories of teacher and student. Interpretations that we offer as a researcher need to be oriented to by the participants if we are to make claims that our interpretations are significant or important (Ingram 2018). Learning is then examined 'in situ and in vivo' (Eskildsen & Majlesi 2018) through students' orientations to the 'learnable' in interaction, and the processes involved can be studied and understood through what students do in interaction.

Another key principle which marks ethnomethodology out from other approaches is the treatment of norms as being reflexively constituted in the interaction itself. It is not that we recognize a particular context as having particular norms that apply, but that we construct and re-construct these norms as we interact, and thus are also constructing the interactional context. Norms are not something that we are obliged to follow; instead they are something we use to help us interact in a way that others can understand, but also to understand the actions of others in this interaction. From a CA perspective, norms are an 'accountable moral choice' (Heritage 1984, 76). Within classroom interaction, teachers and students pursue their goals and intentions through the design of their turns at talk whilst negotiating their 'definitions of the situation' (Blumer 1969) but they also demonstrate the accountability of their participation in interaction through this same turn design (Krummheuer 1995).

One other tool that has been useful in the analysis of classroom interaction is Goffman's idea of a participation framework (Goffman 1981). This describes the 'configuration of all participants…relative to a present speaker's talk' (M. H. Goodwin 1999). Participants can hold different roles within interaction, including the roles of speaker and listener, which are oriented to by the participations themselves and can therefore give us insight into how people position themselves in interaction. Speakers can be 'animators' when they are actually talking, but this may be distinct from the 'author' who is responsible for the words spoken by the 'animator'. Listeners can be 'addressed' or 'unaddressed', or 'eavesdroppers' or 'bystanders'. Within mathematics education this idea has been used by O'Connor and Michaels (1996) to examine 'the ways that teachers purposefully and strategically use language to create contexts for intellectual socialization' (p. 66) and by Krummheuer (2007) who uses the speaker roles of author, relayer, ghostee, and spokesman in his *analysis of participation* in argumentation.

As you will have noticed, CA approaches contexts in a different way to other approaches to discourse analysis, which is fundamental to an institutionally situated research, and we explore the CA conceptualization of context next in order to make sense of how we can study mathematics classroom interactions.

Context

Context has an extremely important role within a CA analysis, but in a very different way to other approaches to analysing classroom interaction. All interactions occur within a context, and these interactions are context-bounded. This context-boundedness of all interactions is referred to as *indexicality*. Within CA, context is not just the environment in which the interaction occurs but is also the interaction itself. It is something that is co-constructed in the interaction by the participants. We show in how we inter-act what aspects of context we are orienting to in the moment. Interactions are thus indexical and reflexive; they are context-shaped and context-renewing (Seedhouse 1996). We cannot explicitly say everything that is needed to make sense of an interaction; a lot is left unsaid, but the property of indexicality allows what is said to represent far more than what is actually said. This also means that we can study the influence of contexts through studying the interactions themselves.

This indexicality and reflexivity leads to a study of sequences of utterances or turns, rather than isolated ones. Each utterance and each turn is shaped by what came before, but also shapes what can come next. There is what Enfield calls an 'enchronic perspective' (Enfield 2013) where analysis is focused on the 'move-by-move flow of interaction' (Enfield & Sidnell 2014, 93). Each turn illustrates our understanding of the context in which we are interacting. We cannot make sense of an individual turn at talk without considering both what came before, i.e. what prompted the turn, and the turns that follow: the effect the turn has including what the turn is doing. This 'next-turn proof procedure' describes how researchers can use this reflexivity to analyse how teachers and students orient to the pedagogical goal within the patterns of interaction. A structure of interaction has 'as a by-product of its design a proof procedure for the analysis' (Sacks, Schegloff, & Jefferson 1974, 728). This next-turn proof procedure enables analysts to examine participants' own methods of achieving and maintaining intersubjectivity (Eskildsen & Majlesi 2018).

So, from a CA perspective, context does not just apply to the institutional contexts or the specific situation in which an interaction occurs, including who the participants are, but it also applies to the interactional context in which each turn is situated. A particular mathematics classroom context is a meso context from this perspective. Within each classroom there are multiple interactional contexts that are influenced by the different goals and intentions in play alongside each participant's turn design.

This view of context particularly influences the analysis of interactions where the focus is on identities or roles. When we observe classroom interactions we can see that the interaction is taking place in a classroom, we can see who is the teacher and who are the students, we can also often see the gender and race of the different participants. There will also be a great deal that we cannot see, such as the sexuality of the participants, whether the teacher is a parent, the qualifications of the teacher, the prior attainment of the students, or the hobbies and interests of the students. The question then arises as to which of these contextual features are relevant to the analysis. In a CA approach, the contextual features that are relevant are those that the participants treat as relevant, not the researcher. This does not mean that the participants have to mention the feature explicitly in the interaction; how they interact as well as what they say can reveal which features of the context they are orienting to. I develop this approach to identity further in Chapter 6.

This approach to considering context does rely on the researcher having some membership knowledge (ten Have 2007) in order to make sense of what participants are doing in the interactions. It is not possible to separate ourselves from our background influences, but it is also not necessarily desirable. As participants we can draw upon a wide variety of background influences and aspects of the context within which we are interacting. These contexts will be viewed differently by those we are interacting with. As researchers, we need to draw upon our knowledge and understanding of the interactional context to interpret how the participants themselves are orienting to this context, whilst still ensuring that it is the participants' perspectives that are used, rather than our own. The analysis presented within this book, particularly in Chapters 5 and 6, is influenced by my own membership knowledge of mathematics classrooms. My experiences first as a mathematics teacher, and then later as a mathematics teacher educator and researcher. As a mathematics teacher educator, I have observed hundreds of mathematics lessons over the years, taught by hundreds of different teachers and student teachers. These observations have enabled me to see similarities and differences in the structures of interactions across a wider range of classrooms, which have inspired me to look more closely at aspects of these structures within my research data.

Conversation Analysis and the Classroom

Conversation Analysis is an established approach to analysing classroom interactions, though it has largely been used in language classrooms rather than mathematics classrooms. As an institutional context, the participants orient to some 'core goal, task or identity (or set of them) conventionally associated with the institution in question' (Drew & Heritage 1992, 22). In the case of the lessons studied in this book, there are at least two goals: that the teacher will teach mathematics and that the students will learn some mathematics. Yet individual classrooms may have other related goals, such as to support their students to explain mathematics, for the students to remember the ideas or procedures they are being taught, or for the students to discover patterns within mathematics for themselves. As the pedagogical goal changes, the structures of interaction, such as turn-taking, also change. This variation in the structures of interaction is dealt with in the following chapters. Chapter 3 focuses on the structure of turn-taking in classrooms and how the variations of this affects the teaching and learning of mathematics. Chapter 4 focuses on the structure of repair and the handling of trouble in interaction. Chapter 5 focuses on how epistemic issues, such as thinking and understanding, are dealt with in mathematics classrooms. It is this relationship between pedagogy and interaction that is of interest to CA researchers examining classroom interaction. Finally, Chapter 6 focuses on what it means to do mathematics in interaction. It describes problem solving, communicating, and argumentation as they are done in classroom interactions.

A key topic of interest in classroom interaction research is learning, as well as other cognitive processes such as thinking, understanding, and knowing. Conversation Analysis approaches these topics in a markedly different way to most educational research. It does not attempt to make any claims about what is going on in teachers' or students' heads: what students do or do not know, or what they do or do not understand. It is interested in how teachers and students interpret each other's actions, treat each other as knowledgeable or not, and establish what it means to do mathematics, as well as how they treat understanding or thinking as interactional objects, all through how they interact with each other. It is epistemological rather than ontological, in that it is interested in the nature of knowledge, the nature of mathematics, and the nature of learning as it is co-constructed in interaction. These topics are often the focus of discursive psychology (Edwards & Potter 1992) approaches to classroom interaction which use the methodology and principles of CA to specifically examine topics that are traditionally the focus of psychology research. Whilst I have not taken a discursive psychology perspective in my own analysis of knowing, understanding, or identities, the analysis does use similar principles and approaches.

Discursive Psychology

Before looking at the relationship between interactional structures and the teaching and learning of mathematics that follows in subsequent chapters, I want to quickly outline one further approach to analysing classroom inter- action that also draws upon all the ideas discussed so far. Discursive psych- ology is an approach within social psychology which makes use of conversation analytic methodology to examine how constructs typically stud- ied by psychologists are constructed, made relevant and oriented to in inter- actions. These include knowing, remembering, or feeling, but with a focus on how these mental states have meaning to the participants in an interaction (Wetherell 2007). Analysis using a discursive psychology approach focuses on how participants construct and use descriptions of these mental states, not what is going on inside the participant's head.

There are several approaches that go by the name of discursive psychology, but here I will focus on the approach first described by Edwards and Potter (1992) which maintains the ethnomethodological underlying principles shared with CA and draws upon many of the methodological tools developed within CA. However, the interested reader will also find the work of Rom Harré and Luk van Langenhove (1999) on positioning theory and the work by Margaret Wetherell (2007), who developed critical discursive psychology, relevant and useful.

Research using discursive psychology examines how participants do 'remembering' or 'understanding' through their utterances, but can also be used to look at identities and how different participants are positioned as learners, knowledgeable, expert, etc. Analysis focuses on how people use descriptions of mental states such as thinking or knowing to perform social actions. These mental states are central to a great deal of educational research since the goal of education concerns students' knowing, thinking, and under- standing. Where CA and discursive psychology differ is that CA examines how issues of knowing, thinking, or understanding are treated in interactions, whereas discursive psychology focuses more on the epistemological nature of that knowledge or understanding. Chapter 5 focuses specifically on knowledge, understanding, and thinking and how it is treated in classroom interactions, and will examine this distinction in more depth.

Conversation Analysis as a Methodology

Interactions are at the same time systematically organized and chaotic; they are also tightly bound to the context in which they occur. Thus, we need to study interaction in its natural habitat. If we want to identify and examine the structures that help students to make sense of what is going on in the classroom, then we need to look at these structures as they naturally arise in classroom interactions. This means collecting naturally occurring data: interactions that would have occurred with or without the audio or video recorder being present. That generally means that the researcher avoids influencing the interaction in any way, though there is some dispute as to whether this means that interactions that result from a classroom intervention can be studied or not. Whilst an intervention may involve the use of particular tasks or talk moves (Michaels & O'Connor 2015), interventions generally do not script what teachers say or how they say things, and it is impossible to script students' responses. Thus, whilst the interactional context may differ from other lessons where there is no intervention, I would argue that this interactional context, and its similarities and differences within 'ordinary' classroom interaction, is worthy of investigation using a EMCA approach. In all classroom-based interactional contexts, this naturally occurring data can appear disorganized, chaotic, and complex, but it is features such as false starts, hesitations, and other markers that can tell us a great deal about the actions participants are performing (Wooffitt 2005).

Conversation analysts generally work with very detailed transcripts from video or audio recordings. Gail Jefferson developed a transcript system specifically for CA analysis (Jefferson 2004). This transcript system includes details of both what was said and how it was said. There is notation for emphasis, changes in speed, changes in pitch, and so on. Some researchers also include gestural information (e.g. Ingram 2014) and aspects such as eye gaze, and there are established transcription systems for each of these. Whilst transcripts are generally the focus of any analysis, the video recordings and audio recordings will be frequently returned to during the process of analysis.

When we come to analyse data, we need to start with the assumption that no detail can be dismissed as irrelevant before we begin. CA analysis is data-driven or inductive. What is of interest is what the participants treat as

```
54   Tyler:    yea:h? (.) but ↑what fractio::n (0.8) what
                >fraction of that< triangle have I shaded
                (0.8) what fraction of that triangle have I
                actually sha::ded (.) Simon,
55   Simon:    um (.) w- ↑half?
56             (0.6)
57   Tyler:    >have I shaded a ↑ha::lf
```

Extract 2-1 Identifying teachers and students in interaction.

relevant in the interaction, not what we as researchers assume to be relevant. This is particularly important when looking at issues such as power, gender, class, etc. We cannot assume these are relevant to the interaction unless the participants themselves are orienting to them. This might be through what is said but could also be through how things are said. For example, in classrooms there are different rights to speak for the teacher and the students. Thus, without anyone being referred to as a teacher or student, it is often quite easy to see who the teacher is and who the students are through the structure of turn-taking. If we look at Extract 2-1, Tyler nominates Simon to take the next turn in line 59, yet Simon does not nominate anyone in the turn that follows and the turn returns to Tyler. This pattern of turn allocation is common to classroom interactions. So, it is not that CA does not consider the background of the participants or the contexts of the interaction, only that it only talks about them if the participants themselves make them relevant.

Teachers generally also control the topic of the interaction and in most classrooms, but not all, say more than the students. These are both features of Extract 2-1, where Tyler speaks for longer than Simon and in full sentences, but also controls the topic of the interaction through his question. Simon, in contrast, only speaks to answer Tyler's question using just one word, but also gives this answer hesitantly. Each of these are not necessarily issues or features to change, as a lot depends on what is being said and how the topic is controlled; on the contrary, they may enable students to make sense of what is going on in the classroom and support them to learn.

Transcription

In this section I focus on transcription, as this is emphasized so much in CA approaches and is a key feature of later chapters. In CA, analysis involves focusing both on the video or audio recordings of the data and the transcriptions of these recordings. However, in the reporting of data there is still an emphasis on the sharing of extracts from transcriptions only, particularly

[	left square bracket	indicates the point of overlap onset
[	right square bracket	indicates the point at which two overlapping utterances end
=	equal sign	indicates no gap
(0.0)	numbers in parentheses	indicates elapse time measured in tenths of a second
(.)	a dot in parentheses	indicates a brief interval of less than 0.3 seconds
.	a full stop	indicates a falling, stopping intonation
,	comma	indicates that a speaker has not finished, usually marked by a fall-rise or weak rising intonation.
?	question mark	indicates a rising intonation often associated with questioning.

Figure 2-1 Transcription conventions related to turn taking.

when data involves children, such as in the study of classroom interaction, where ethical considerations and consent often restrict the data that can be shared. In CA, transcripts are not data in and of themselves but are part of the analysis and a representation of the data (Ingram & Elliott 2019). The aim is for the transcript to include as much detail as possible, so they include features such as the duration of pauses, prosody, audible breaths, laughter, where words are stressed, elongated, and truncated, and the details of where talk overlaps (Hepburn & Bolden 2013). However, this level of detail is often unreadable by someone less familiar with CA transcription. CA has a distinctive style of transcription developed by Gail Jefferson (2004) which is common across almost all CA research and therefore becomes more familiar and easier to read over time. In this book I have followed the principle of only including the transcript details relevant to the analysis presented, rather than all the details included in the transcript that I worked on at the time. This, I hope, will make the extracts more readable by people less familiar with CA. Drawing on Sidnell's (2010) descriptions of the Jefferson transcription system, I will group the transcription notations by the role they have in the reading and interpretation of what is being said. Several features of the Jefferson transcription system focus on aspects of talk that relate to turn-taking as shown in Figure 2-1.

Each of these symbols represents an aspect of talk-in-interaction that shows us when turns end or change, or where a transition relevant place may be. Notice that the punctuation symbols have specific uses related to intonation rather than to the grammar of what is being said. As the precise location of overlapping speech is also important, Jefferson transcription also uses an equal width font: Courier New. Other symbols mark features of talk that relate to how something is being said as shown in Figure 2-2.

word	underlining	indicates emphasis which is located within a word
WORD	capital letters	indicate speech that is hearably louder than surrounding speech
°word°	degree signs	indicate speech that is hearably quieter than surrounding speech
wo::rd	colon	indicates the degree of elongation of the prior sound, the more colons the more elongation
wo-	hyphen	indicates a cut-off of the preceding sound
>word<	inward arrows	indicate speech that is hearably faster than the surrounding speech
<word>	outward arrows	indicate speech that is hearably slower than the surrounding speech
↑word	upward arrow	indicates onset of a noticeable rise in pitch
↓word	downward arrow	indicates onset of a noticeable fall in pitch

Figure 2-2 Transcript conventions for how talk is spoken.

→	right arrow	indicates a significant line that is being discussed in the analysis
(words)	single brackets	indicates a guess of what is being said as it is unclear
( )	single brackets	indicate unclear talk
((cough))	double brackets	transcriber notes to represent something this is hard to write phonetically

Figure 2-3 Transcriber annotations.

There are also a few symbols that are used by the transcriber for transcription purposes rather than to indicate features of the talk being transcribed (Figure 2-3).

Each of these symbols was used in the transcription of the whole-class interactions used in the analyses presented in this book. There are additional symbols that have been more recently developed to deal with gestural information, but these will be dealt with as they arise in this book. There is a level of interpretation involved in the transcription of classroom data. For example, some of the teachers videoed had strong regional accents that influenced how parts of a sentence were emphasized or affected the intonation of words. The same is true of many of the students, some of whom had English as an additional language, which can also affect the pronunciation of particular words. I have used the principle of transcribing aspects that were audibly different to the surrounding talk, but the transcription features included are part of the analysis of the data. For some parts, I have used software such as Audacity˚ (2018) and Praat (Boersma & Weenink 2018) to more accurately record details of the interaction, such as the length of pauses.

Patterns and Structures of Interaction

There are structures that underlie how we interact with each other, which mean we can make assumptions and adopt norms without speaking about them. Conversation Analysis has largely focused on the structures of turn-taking, adjacency pairs, preference organization, and repair. These structures are referred to as rules, norms, and routines by different authors (and are similar to the idea of practices in sociocultural research). They are generally implicit but are more noticeable when things go wrong in an interaction. We notice when someone says something when they shouldn't, we notice when someone doesn't answer a question we've just asked them, and so forth. However, these rules are not prescriptive, in fact we can use them to demonstrate different positions towards the other speakers (Harré & Van Langenhove 1999), but they do enable to us to interact and make sense of what other people are saying when they interact with us. It is these structures that CA researchers are interested in. The aim of the analysis is to identify and describe the organization or order that is co-constructed in interactions, such as those co-constructed by teachers and students in the mathematics classroom.

Deviant case analysis is another key feature of CA research. If an interactional sequence does not follow the usual pattern or structure identified by the research, this is not considered irrelevant but instead as highly informative. It is the detailed analysis of these deviant cases and how they deviate from these patterns that tells us more about how these patterns are generated, used, and made relevant. The patterns and structures within interactions are made visible when they are deviated from, as participants need to account for this deviation (Tsui 1991). We return to this idea in Chapter 4 when we examine the treatment of student errors.

I will now turn my attention to four interactional structures, described by Sacks and his colleagues, that are widely used by researchers in CA, and I draw upon each in my analysis in later chapters. These are structures that participants use in their interactions to both make sense of what the other participants are doing, and also to enable others to make sense of them. Some of these structures are different in classrooms from how they are in ordinary conversations, and this can tell us a great deal about the nature of classrooms. In other cases, the structures remain the same, but how teachers and students use them differs from how they are used elsewhere.

Turn-taking

'the term "turn" or "turn at talk"... refers to an opportunity to hold the floor, not what is said while holding it' (Goffman 1976, 271)

Turn-taking is one area where classrooms differ noticeably from ordinary conversation. The structure of turn-taking in ordinary conversation is designed so that it minimizes overlaps and gaps between speakers. In contrast, turn-taking in classrooms is structured in such a way as to allow for gaps between speakers; this is explored further in Chapter 3. The structure for turn-taking in ordinary conversation is detailed at some length in a paper by Sacks and colleagues (Sacks et al. 1974) and is adaptable to the context in which an interaction occurs. Interaction generally runs quite smoothly, with speakers 'knowing' when to speak and not speak, even though no two turns at talk are alike. Turns can vary in length and structure, they can be one word long or several sentences long, yet we are able to judge when it is appropriate to speak and when it is not. Conversation Analysis uses the notion of a *transition relevance place* (TRP) to indicate the point where a change in speaker may occur. It is at these points that the rules or structures of turn-taking come into play. If the current speaker selects the next speaker during their turn, then this selected person must speak next. If the current speaker does not select the next speaker, then at a TRP any other participant can speak, and the person who speaks first takes the turn. However, if no one speaks at this point, then the turn returns to the original speaker.

The structure of turn-taking in classrooms is detailed by McHoul (1978) and reflects the different roles of the teacher and the students. If the teacher is speaking, then the teacher can nominate the next speaker, and this student must speak next. If the teacher does not nominate a student to do this, then it is the teacher who continues to speak. This is the first place where the structure differs from ordinary conversation, as students can only take the turn if the teacher nominates them. If it is a student who is speaking, then this student can nominate the next speaker. If they do not nominate the next speaker, then the turn becomes the teacher's. Only if the teacher does not take this turn do the other students get the opportunity to speak. This turn-taking structure is illustrated in Figure 2-4.

This structure for turn-taking in the classroom does not allow students to self-select as speakers, but it does allow for gaps between turns, as there is not

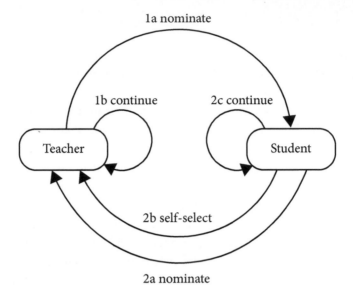

Figure 2-4 The structure of turn-taking in most classrooms.

(Reprinted from Ingram, J., & Elliott, V. (2016). A critical analysis of the role of wait time in classroom interactions and the effects on student and teacher interactional behaviours. Cambridge Journal of Education, 46(1), 1–17 reprinted by permission of the publisher (Taylor & Francis Ltd, http://www. tandfonline.com)).

competition to be the first speaker like there is in ordinary conversation. This is an aspect of the structure that teachers can use to their advantage, as I discuss in Chapter 3. These rules give a great deal of control over who can speak to the teacher, and for how long. Whilst there has been substantial work trying to adapt these structures, and it has been criticized because of the asymmetry in the rights and responsibilities of teachers and students, the structure does enable students to take time to think, and gives teachers the ability to control the topic of interaction, so it is more how the structure is used than the structure itself that is at issue.

A variety of structures of turn-taking operate within a classroom. The structure described by McHoul dominates formal teacher–student interactions, such as the majority of whole-class interactions, but different structures more similar to ordinary conversation have a role when students are working in small groups or pairs, for example.

Adjacency Pairs

Adjacency pairs are the simplest unit that is considered in CA, or, as Heritage describes them, 'the basic building-blocks of intersubjectivity' (1984, 256). In their simplest form they include two parts: a first-pair-part (FPP) spoken by the first speaker, and a second-pair-part (SPP) spoken by another speaker. Common ones include questions and answers, requests and acceptances, invitations and acceptances.

Adjacency pairs include the idea of *conditional relevance*, which will also appear below when I talk about preference organization. If someone asks a question, then an answer is conditionally relevant in the next turn. If an answer is not given, then it remains relevant but is 'noticeably absent' (Bilmes 1988) and accountable. This does not mean that an answer will always follow a question, but that it is a norm that enables us to draw conclusions about the person we are interacting with when they do not give the expected answer, or from the account they offer for why they have not answered the question.

The FPP of an adjacency pair can be analysed as projecting a specific next turn; a question projects that the next turn should be an answer. If the relevant SPP—an answer, for example—is given, it will usually be treated as seen-but-unnoticed. However, if what follows is not the relevant SPP, then its absence is noticeable and accountable. If no account is given, as well as no relevant SPP, then the turn is sanctionable by the first speaker. This structure is analytically useful, as we can examine the role that different SPPs and different accounts have in interactions. In particular, accounts are used by students to explain their answers, both when they can't give the answer the teacher is looking for and when they have given an answer that is not expected, for example at different points in the interaction. This is explored further in Chapter 4.

One pattern of classroom interaction that is well documented is the IRE or IRF pattern (Mehan 1979a; Sinclair & Coulthard 1975), which includes a question from the teacher, the initiation (I), a response from a student (R), and then an evaluation (E), feedback (F), or follow-up move (F) by the teacher. The FPP is the question asked by the teacher, and the relevant SPP is an answer from the student. The third turn displays an analysis of the student's response, and from this turn the student is able to interpret their answer as suitable or not. This is not only done through an explicit evaluation, but there are a variety of third-place moves a teacher can make that can indicate whether the student's responses was appropriate (Lee 2007). For example, if a teacher moves the interaction on without comment, this implies there was no issue with the student's response. Further exploration of this IRE structure is considered in Chapter 3.

This relationship between turns at talk is used within CA as a form of validation for the analysis. The second turn displays an interpretation of the first turn, which in turns creates an action for interpretation in the turn that follows. Participants' understanding of each other develops as this sequence of turns develops. This relationship is referred to as the *next-turn proof procedure* (Sacks et al. 1974, 729 and as discussed above). The turns themselves document the analysis of previous turns and display this analysis in the way the turn is constructed. In interaction this is what enables participants to make sense of what is being done, but it is also available to us as researchers. Hence CA analyses sequences of interactions using data which is publicly available and is available to be shared with readers of this research.

Preference Organization

Preference organization refers to the phenomenon that some turns are preferred over others. It is used in different ways by researchers and can be problematic, as a turn being preferred does not mean that it is liked or wanted. It is about affiliative or disaffiliative actions, it is about whether actions are sanctionable or accountable or not. A preferred action is one that follows the social norms; if you ask a question, the preferred response is an answer to that question. An answer is neither sanctionable nor accountable. If an answer is not given, then an account for why not is expected, and if that does not happen either, then the turn is sanctionable. Actions are affiliative if they allow the interaction to proceed smoothly, and the structures of interaction enable this to happen (Stivers & Robinson 2006).

For many adjacency pairs there are alternative SPPs, and some of these are preferred over others. These alternatives are performed in different ways. Preferred actions are normally given in an unmarked way, without hesitation (Pomerantz & Heritage 2013). On the other hand, dispreferred responses are marked by hesitations, delays, and/or accounts. Dispreferred responses are usually mitigated in some way to minimize their disaffiliative effect. Particular types of response are not necessarily preferred just because of their type. An agreement is not always preferred; for example, when someone makes a self-deprecating comment, agreement is dispreferred (Pomerantz 1984). Furthermore, dispreferred actions are not always sanctionable. If a dispreferred response is given in such a way that its disaffiliative nature is minimized, then it is no longer sanctionable. In classrooms a preferred response is often considered to be one that is used by the teacher in a way that contributes to the goal of the lesson (Greenleaf & Freedman 1993).

Returning to the idea of interactional structure, a preferred SPP is seen-but-unnoticed and given immediately, enabling the interaction to continue smoothly. Dispreferred responses are given more hesitantly, but this hesitation allows the person responding more time to devise an account and ways to mitigate the dispreferred response. In classrooms students also use hesitation markers such as 'um' to buy time to think through their answer to a teacher's question, and not necessarily to mitigate a dispreferred response.

Preference organization does not just refer to the structural bias in alternative SPPs of adjacency pairs, it also applies to other interactional structures such as that of repair, described below. This is another interactional structure that is highly relevant to classrooms, as it deals with sources of trouble in interactions, including mistakes. We also return to this idea of preference in Chapters 3, 4, and 6.

Repair

The organization of repair is another key theme explored within CA research. Repair describes the way in which trouble within an interaction is dealt with. This could be trouble in terms of someone mishearing or misunderstanding what is being said, or it could be a mistake, a dispreferred response, or speaking out of turn. Analytically, a repair is split into three parts: the trouble source, the initiation of the repair, and the performance of the repair; whose turn these take place in is also of interest. Self-repair initiation or performance is when the person whose turn contained the trouble source initiates or performs the repair, whereas other-initiated repair or performance is when it is done by someone else. Schegloff, Jefferson, and Sacks (1977) showed that the interactional structure of repairs enabled a preference for self-correction in ordinary conversation. They showed that self-initiated self-repairs in the same turn are the most frequent type of repair, with other-initiated self-repair and other-initiated other-repair occurring very rarely. This preference for self-initiated self-repair is part of the structure of interactions in that the trouble source, initiation, and repair often occur within the same turn. For example, in Extract 2-2, Tim begins to asks a question which he cuts off at m-, and self-repairs to asking why before completing his question about what the

```
289 Tim:        right any offers anyone for telling me what,
                m- why of course we always want to know why
                (.) what the mode, the median the mean and
                the range are (1.7)
```

Extract 2-2 An example of self-repair.

```
330 Tim:        what was the lowest number of days absent.
331 Steve:      zero or one I don't know
332 Tim:        you don't know. ok someone else then, what's
                the lowest number of days pe- someone was
                absent.
```

Extract 2-3 Another student performs a repair.

mean, mode, median, and range are. This all happens within a single turn, with no gap between the point that Tim treated as the source of trouble and his initiation and performance of the repair.

Furthermore, there is often a delay before other-initiations of a repair, which offers a further opportunity for self-initiation and self-repair before the turn is taken by someone else. The fact that other-initiated self-repairs are far more common than other-initiated other-repairs also shows how other people orient to this preference for self-repair, as the initiation of the repair allows self-repair even though the other initiating the repair could also have performed the repair themselves.

However, Schegloff and colleagues did note that other-repairs were more common in adult–child interactions as 'a device for dealing with those who are still learning or being taught to operate with a system which requires, for its routine operation, that they be adequate self-monitors and self-correctors as a condition of competence' (1977, 381). It therefore seems reasonable that other-initiations and other-repairs will be more common in teacher–student interactions, as many researchers have shown (Liebscher & Dailey-O'Cain 2003; Macbeth 2004; McHoul 1990). However, this more frequent occurrence of other-repair should be transitional as students become more competent and are able to both self-initiate and self-repair.

In classrooms it can also be useful to distinguish between teacher-initiated and peer-initiated or performed repairs when the trouble sources occur in a student's turn, as explored in Chapter 3. Thus, there are two types of other-initiation and two types of other-repair. Other-repairs are far more common in classroom interactions than in ordinary conversation. For example, in Extract 2-3 Steve initiates a self-repair in his own turn in line 331, but Tim invites another student to perform the other-repair.

Additional work focusing on classroom interactions has looked specifically at the type of other-initiation. Liebscher and Dailey-O'Cain (2003) distinguish between how specific the initiation is, with words such as 'pardon' or 'what' giving no indication of what the trouble is, question words such as 'who', 'what', or 'when' which give some information as to what the trouble is, as well as full or partial repeats which usually give a more specific indication of what the trouble is. These types of specific repair initiations are far more common in classrooms than in ordinary conversation, particularly when students initiate a repair.

At a methodological level, these features of interaction, turn-taking, preference organization, and repair are analytic concepts. The aim of this chapter was to introduce you to the principles that underpin the analysis that follows in the rest of this book, but also to introduce you to the structures of interaction that are well documented in research, both in ordinary conversation and in classroom interactions. These structures help us to make sense of how interaction in the mathematics classroom can both enable and constrain what mathematics is learnt and how it is learnt. By what mathematics is learnt I am not talking about particular topics or concepts, but instead the nature of mathematics as it is communicated and co-constructed by teachers and students in classrooms.

Conclusion

This chapter has introduced Conversation Analysis as an approach to the study of mathematics classroom interactions. This is the approach I have used throughout my work on classroom discourse and in the analysis of all the data I draw upon for this book. The underlying principles developed out of ethnomethodology and symbolic interactionism are key to understanding the work presented in this book, in particular the focus on how things are said, as well as what has been said. Considering many of the ideas and concepts that are the focus of a great deal of educational research as actions, rather than as objects, enables us to understand more about how classroom interaction can affect the teaching and learning of mathematics. The aim of the following chapters is to illustrate these effects and the consequences of particular interactional structures that are often unnoticed by teachers, students, and researchers.

3

Turn-taking

To most observers, the contrast between ordinary conversation and class-room interaction is most evident in the organization and allocation of turns at talk. Classroom interactions are highly complex, yet they are also struc-tured and ordered. Teachers and students use a range of strategies and resources to achieve this orderliness, which is most clearly evident in the structures of turn-taking. In all interactions there are implicit structures or rules governing who speaks when, how long they can speak for, and what can be said. These structures are made visible through how participants orient to them and through the sanctioning of participants when these structures are deviated from. Sacks et al. (1974) point out that there is a relationship between the turn-taking system and the types of activities that are being undertaken in the interaction. Seedhouse (1996), for example, argues that the IRE (Initiation Response Evaluation) sequence described below is widely used in learning contexts because it is suited to the core goal of education.

In this chapter, I focus on the structure of turn-taking during whole-class interactions in mathematics classrooms. I return to the discussion in Chapter 2 on the structure of turn-taking in whole-class interactions as described by McHoul, before examining interactional contexts where these structures, or rules, are deviated from. It is the study of these deviations that reveals the consequences these structures have on teaching and learn-ing opportunities. One common structure of turn-taking is often described as an IRE sequence, and I also show how discourse analysis and Conversation Analysis differ in the way that they identify and analyse this structure, and the consequences this has when considering the actions teachers and students are performing within this structure of turn-taking. An in-depth analysis of the structure in terms of where pauses can and can-not be left also reveals why wait time has the effect on students' responses that it does.

Patterns in Mathematics Classroom Interaction: A Conversation Analytic approach. Jenni Ingram,
Oxford University Press (2021). © Jenni Ingram. DOI: 10.1093/oso/9780198869313.003.0003

The Structure of Turn-taking in the Mathematics Classroom

In Chapter 2, McHoul's (1978) description of the 'rules' of turn-taking in 'formal' classrooms was outlined. These rules illustrate the constraints within the local management of turn-taking in classrooms. For example, these rules do not allow for students to self-select as the next speaker when the teacher is currently speaking. These restrictions on students self-selecting minimizes the possibility of overlap or interruption in classroom interactions, whilst at the same time increasing the opportunities for gaps or pauses between turns, compared to ordinary conversations. Each of these features is examined in more detail below as we consider how they—and their different uses—may influence teaching and learning. Turn-taking structures are context dependent (Sacks et al. 1974), and within a classroom there will be different structures for whole-class interaction, small group work, and one-to-one conversations between the teacher and a student.

Teachers can allocate these turns at talk by naming the student to speak next, but also through what Käntä (2012) calls 'embodied allocations', such as nodding, pointing, or simply making eye contact. Students can also bid for the next turn by raising their hand, or adjusting their body language to invite or discourage the teacher to nominate them (Fasel Lauzon & Berger 2015). There have been several initiatives over the years to alter these allocation practices, such as the 'no hands up' (Wiliam 2011) rules found in many classrooms, or teachers being encouraged to involve more students to participate by asking students who have not raised their hands (Waring 2014). More recent studies have begun to question the control teachers have over turn allocation, arguing that it 'is achieved actively and collaboratively' (Fasel Lauzon & Berger 2015, p. 15). Koole and Berenst (2008) also suggest that teachers are more likely to choose students who have in the past frequently bid for a turn, and this is further supported by Rowe's research (Rowe 1972) into wait time.

Breaching the Turn-taking Structures of the Classroom

Building on McHoul's (1978) detailed description of the structure of turn-taking in formal classrooms, here I will explore what happens when whole-class interaction deviates from this structure. Teachers generally visibly dominate turn allocation—that is, they control who gets to speak, what they

```
60  Tanya:    okay, thank you. Simon
61  Simon:    with the shape look it shows that the
              quarters of like (.) that how much of that
              like, how much area was ((inaudible)) the
              four quarters ((inaudible)) quarters
62  Tanya:    right so maybe (.) maybe they could have used
              the shape a bit (.) more with the (.) given
              that there was four of it and it was about
              dividing by four. four edges sorry. yeah okay
              right, hands up if you understood what was on
              that table. (.) okay a few of you. Stella
              could you have a go, just bring it to the
              front. hands up if you didn't understand.
```

Extract 3-1 Teacher controls turn-taking.

get to say or do, and how long they get to speak for (though students are able to influence these processes (Fasel Lauzon & Berger 2015))—and it is the structure of turn-taking that enables them to do this. This is even the case when the interaction is focusing on what the students are thinking, and their own ideas or understanding of something, rather than the ideas of the teacher. In Extract 3-1 Tanya begins by acknowledging what the previous student has said in turns 57 and 59 (omitted from transcript), before nominating the next student to speak. The students are all responding to an invitation by Tanya to share their thoughts on the problem they had been working on since the beginning of the lesson back in turn 26. Tanya nominates which students share their thoughts on the problem, as she does in turn 60, but the topic of the interaction is controlled to some extent by the students as they were reporting the different ways in which they thought about the problem. After each student has spoken Tanya always takes the next turn, as she does in turn 62, often by repeating or rephrasing what the student has said in their turn before.

This sequence of teacher selecting next speaker, student responding, teacher taking the next turn is often referred to as the IRE sequence, and there has been extensive research into the uses and abuses of this structure in a range of classrooms, which I explore later in this chapter. However, to begin with I will explore situations in whole-class interactions where there are deviations from the usual structure of whole-class interactions that was outlined by McHoul (and in Chapter 2) and these are not sanctioned. In McHoul's rules, students do not have the opportunity to self-select as the next speaker, and McHoul reports that this does not happen in his data. Rules are deviated from all the time in interaction, but participants are usually sanctioned or held accountable for these deviations in some way. So, these situations where the rules are deviated from, and there are no sanctions, tells us that something else is at play that overrides the normal structure of turn-taking.

There are three interactional contexts where there are examples of students self-selecting as the next speaker. The first is to ask a question, the second is to initiate or perform a repair, and the final context is where the teacher has not nominated the next speaker in the slot where this usually happens (Ingram 2010). Each of these contexts is very rare compared to the occasions where the rules are oriented to, but it is this deviation from the norm that makes these occasions analytically interesting. They are deviant cases. These contexts also all illustrate the pedagogical purpose of classroom interactions, as in each case the deviation enables the teacher to continue the interaction, developing the mathematical activity that is the focus of the interaction. This phenomenon is referred to as the preference of progressivity in interaction (Stivers & Robinson 2006) within CA.

When Students Ask Questions

Turning to the first scenario, where a student can take a turn that does not follow the usual structure of turn-taking, we consider here the situation where a student is asking a question. In classrooms many of these questions are about the tasks and what students need to do, but occasionally a student can ask a question about the mathematics. For example, in Extract 3-2, Tyler is explaining how to calculate a probability and a student asks 'how' before Tyler has completed his explanation.

Tyler repeats this question from Seb which also serves to clarify which numbers Seb is referring to by the word 'that', which appears twice in his question. There is then a pause of 1.1 seconds before Tyler begins to answer the question. In this scenario the student has self-selected at a point where he could be considered as interrupting the teacher, though he does not overlap

```
234 Tyler:    one in eight. ok. if I cancel them down, that
              and that cancels. that and that cancels I'm
              left with (0.7) a tenth. so-
235 Seb:      how do you know that cancels with that
236 Tyler:    how do you know that this cancels down
237 Seb:      yeh
              (1.1)
238 Tyler:    if I multiplied it out you'd see tha-, that
              (0.3) I have a factor of eight on the top and
              a factor of eight on the bottom.
```

Extract 3-2 Example of a student self-selecting to ask a question.

with Tyler. Here Tyler acknowledges and responds to Seb's question and does not sanction him for speaking out of turn.

Again, this can be argued to be part of the preference of progressivity, as it enables Tyler to support his students to understand the explanation he is giving. Tyler is explaining a process for simplifying a multiplication of several fractions by cancelling common terms in the numerator and denominator, and Seb has indicated that there is something not clear about this explanation. The pedagogical focus of this extract is the explanation of the simplification process which builds towards the larger focus of the interaction of noticing that the probability of each event, including the one being calculated in Extract 3-2, is 1/10. These questions help to maximize the effectiveness of the teacher's explanation.

The limited opportunities for students to self-select within the normal structure of turn-taking in whole-class interactions also explains why students asking questions is so rare in whole-class interactions, yet, as shown in Extract 3-2, students are generally not sanctioned for asking a question.

One way in which students self-select is by calling 'miss' or 'sir'. Yet here the student is only asking for permission to speak at this point; they have not said what they want to say, or asked the question they wanted to ask. This asking of permission is one way that students can mitigate the dispreferred nature of self-selecting to speak.

When Students Initiate a Repair

The topic of repair is dealt with more extensively in Chapter 4, but essentially it is about the treatment of interactional difficulties or sources of trouble. Mistakes are just one source of trouble in interactions, but are often problematic in terms of the pedagogical focus of an interaction. In Extract 3-3 the interaction is between Tyler and Seth. In turn 32, Seth gives the incorrect

```
31  Tyler:    one in ni:ne (.) so has that gone up or gone
                down
32  Seth:     gone down
33  Tyler:    probability's gone
34  Seth:     u[p ]
35  Sasha:     [up]
36  Sonia:    up
37  Tyler:    probability's gone up, it's more likely now
                that you're going (.) to get (0.3) the (.)
                red cross (.) so Sid (.) choose one
```

Extract 3-3 Students self-select to repair an earlier response.

response to Tyler's alternative question in turn 31, a question that makes relevant an answer that is 'one of the two alternatives in the question' (Koshik 2005).

Even though Seth self-corrects and gives the acceptable response in turn 34, as shown by Tyler's acceptance of it in his repetition in turn 37, Sasha and Sonia both self-select to give the correct answer as well and consequently repair the trouble. Again, this new answer of 'up' allows the teacher to continue with activity. This example also illustrates how the pedagogical purpose of the interaction drives the structure, which in this instance does not involve dwelling on the mistake made or exploring the class's knowledge further. In Chapters 4 and 5, I return to the different roles that errors play in interactions concerning the treatment of mathematical knowledge. The different ways that teachers treat errors in interaction influence the status and nature of knowledge within these interactions.

When the Teacher Does Not Nominate Someone to Answer

The final context is illustrated in Extract 3-4 and Extract 3-5. In Extract 3-4 Tyler asks a question but does not select a specific student to answer the question. Instead, three different students all offer an answer, in different forms, which Tyler accepts in turn 138.

Similarly, later in the same interaction (Extract 3-5), Tyler asks another question in turn 176 and again does not select the next speaker. This time two different students answer but with contrasting answers in turns 177 and 178. This happens again where Tyler re-asks the question but adds a visual prompt in turn 179, and two other students again give contrasting responses in turns 180 and 181. In turn 183 the student self-selects whilst Tyler is still speaking and has not given Tyler the opportunity to select the next speaker. This extract suggests that for some interactional contexts the norms of turn-taking differ from those described by McHoul. In both Extract 3-4 and Extract 3-5 the teacher still has the right to take the next turn without students talking at the

```
134 Tyler:    what's the probability?
135 Steve:    a hal[f     ]
136 Sarah:         [fifty fifty]
137 Simon:         [[a ha]lf  ]
138 Tyler:    fifty fifty, a ha:lf, good
```

Extract 3-4 Several students self-select to take the next turn.

```
176 Tyler:     one. okay. what number would be most likely
                do you think. what total would be most likely
177 Shannon:   [si:x ]
178 Steve:     [eight]
179 Tyler:     what total will go (5.4)
                ((draws diagonal boxes in the grid))
180 Sian:      seve[n ]
181 Sasha:         [si]x
182 Tyler:     what total would go diagonally across the
                [board]
183 Shannon:   [six  ]
184 Sam:       seven
185 Tyler:     seven (.) good. seven would be the most
                likely and there'd be six of those (.) out of
                (.) a total of thirty six, so. how do we ge:t
                (1.8) one out of thirty six quickly if we
                know (.) the probability of getting a six (.)
                is one sixth.
```

Extract 3-5 Students self-select with contrasting answers.

same time, and it is Tyler who ends the interaction by accepting one of the responses.

Tyler can take the next turn without difficulty, and his response shows that the students' answers are what he was expecting. In fact, unison responses and simultaneously given appropriate responses have the same structure. In effect, the teacher is interacting with the class as a whole as if it were a single student (Lerner 2002; Payne & Hustler 1980). This is described by Rowe (1972) as a two-player game conceptualization of the classroom. If we treat the interaction as being between two people, the teacher and the class, then the structure of turn-taking is the same as that for ordinary conversation and for classrooms. There are also occasions where several students speak simultaneously or overlap with each other where they have given different answers. The structure of the interaction can then deviate from that of usual classroom interactions; this is explored further in Chapter 4.

If we treat the interaction as being between the teacher and the class, where teachers treat students as a 'unified cohort' (Payne & Hustler 1980), rather than between the teacher and a specific student, then the norms of turn-taking are still being oriented to. This would also be the case in situations where it was a classroom norm to give choral response. Chanting, choral responses, or other scenarios where multiple students self-select to respond to a teacher's question are becoming more common in some classrooms. These generally happen when the questions are relatively simple and usually involve the students performing routine processes such as arithmetic calculations. In other words, these undirected questions usually occur when the teacher can

```
268 Tyler:    two: ok? which is the first prime number.
269 Shane:    one
270 Sid:      tw[o
271 Salma:       [zero
272 Sid:      two
273 Shane:    one
274 Tyler:    one[two
275 Saul:        [two
276 Sibyl:    is seven one sir?
277 Tyler:    okay we might discuss it tomorrow
278 Sian:     one
279 Tyler:    which one's the first prime number?
280 Pupil:    um
281 Tyler:    why one?
282           (2.1)
283 Tyler:    why is it not a prime [number ]
284 Shane:                          [because] you can only
              divide it by, one
285 Saul:     yeah but you can divide it by itself, because
              it divides by itself
286 Sibyl:    yeah but you can't, you can only divide it by
              one though
287 Saul:     yeah and that's dividing by itself
288 Sibyl:    yeah but one's (((inaudible)) number)
289 Tyler:    wind them up and let them go. so what is it,
              one or two, one or two.
290 Pupil:    t[wo
291 Pupil:     [one
292 Pupil:    two
293 Pupil:    one
294 Pupil:    two
```

Extract 3-6 Students self-select to address each other.

be confident that many of the students will be able to answer and give the same answer. This means that the responses are heard as being convergent (Schegloff 2000) rather than competing for the right to the turn. However, even this structure of turn-taking in teacher–class interaction can be deviated from.

Multiple students self-selecting to answer a question can also result in a further deviation from the rules of turn-taking. In Extract 3-6 the three students, Shane, Sid, and Salma, all give an answer to Tyler's question of 'what is the first prime number?' These three answers are all different. Shane and Sid give their answers simultaneously and Salma gives hers soon after, overlapping with Sid. The interaction then continues between the students, with Tyler re-asking the question in turn 279, then changing the question in turns 281 and 283. As the students continue their interaction but offering reasons for why one is a prime number or why one is not a prime number in turns 284 to 288, there is no overlapping of turns. This is despite the fact that the students are self-selecting and giving different answers. Here the structure of turn-taking more closely resembles that of ordinary conversations rather than formal classrooms as the students interact with each other.

These arguments or debates require a deviation from the norms of classroom turn-taking if they are to happen, as they involve students explaining and justifying their ideas to each other, not the teacher. When classroom interactions do deviate from the usual structure of turn-taking, such as when the turns become between two or more students, the teacher does not have to do much interactional work to take a turn and re-establish the structure.

Sanctioning Speaking out of Turn

Whilst in the examples above no one has been sanctioned or held accountable for deviating from the norms of turn-taking, there are also occasions where the same actions occur and students are sanctioned by the teacher. However, this can simplify the different roles interactions can have in the classroom, and the different structures of interaction that come into play. Classroom interactions have different roles and purposes, and can form different interactional contexts which may result in variations in the structure of turn-taking. Some of these different interactional contexts are considered in more depth in Chapters 5 and 6, but here I focus on the sanctioning of deviations from turn-taking structures.

Sanctions can happen in a variety of ways. For example, a student can speak when the turn is not theirs and the teacher can explicitly sanction this, as in Extract 3-7.

The students can also sanction another student speaking when the turn is not theirs, as illustrated in Extract 3-8.

But the teacher can also ignore what the student has said, so their turn is not evaluated or acknowledged, as in Extract 3-9. This is an implicit sanction embedded within the IRE structure of classroom interactions, as a student's response is usually followed up on by the teacher.

```
29  Trish:     you're going on a lit bit of a tour (0.9) and
               you're in Madrid (0.4) and then you go to
               Moscow (.) what happens then (.) ↑hotter or
               colder and by how much
30             (2.4)
31  Stephen:   twenty-eight
32  Trish:     don't shout out please but Stephen said tell
               me
```

Extract 3-7 Sanctioning for speaking out of turn.

```
235 Trish:    in we've already had j so e thank you Scott
              do you think that's true or false
236 Scott:    true
237 Seth:     true cause there's only three outcomes to get
              so we-[
238 Scott:         [shut up I was  ((inaudible))   here]
239 Stefan:        [shut up so it's gonna be a (third)]
              shut up
240 Trish:    er excuse me. just settle down. could you not
              tell each other to shut up because it's very
              rude
```

Extract 3-8 Students sanctioning each other for speaking out of turn.

```
89  Trish:    somewhere in between. do you think it's
              closer to certain or closer to impossible
90  Sophie:   impossible
91  Stella:   certain
92  Trish:    close to certain okay
```

Extract 3-9 Ignoring what a student has said when self-selecting.

Pedagogical Advantages of the Turn-taking Structures

The turn-taking structure of classrooms has many pedagogical advantages, as well as some disadvantages. The prevalence of, and preference for, teachers nominating the next speaker can be used by teachers to ensure a range of students have the opportunity to speak, that questions are targeted towards particular students, and that what students say can be built on and used. This control has both advantages and disadvantages, depending on how it is used by teachers. It can just as easily be used to ensure that a range of views or opinions from different students, perhaps varying by social status or prior attainment, are heard and considered, as it can also be used to limit the opportunities of these different students, for example by only asking students with prior low attainment very simple factual questions.

The turn-taking structure also gives the teacher more control over the topic that is being discussed. Classroom interactions are influenced by the institutional goal that students need to learn something, and the teacher's goals within this will include specific goals about what mathematics they want the students to understand or do. The turn-taking structure enables the teacher to establish the topic and to maintain the topic of the interaction (van Lier 1988). The teacher can keep their turn even if they pause for a significant amount of time, as no one else in the classroom has the right to speak. In this turn the teacher can introduce a new topic, open up a topic for discussion, or close down a topic. The control over turn allocation and the constraints placed on student responses to teacher initiations also enables teachers to establish the interactional context, which influences the nature of the mathematical activities and behaviours students engage in, and consequently their learning of mathematics. It is largely through the turn-taking structure, including the IRE pattern of interaction discussed below, that the interactional context is established and maintained.

Initiation-Response-Evaluation (IRE)

The term IRE has arisen several times so far in this book, and now I turn to examining this structure of turn-taking in more depth. The IRE sequence is a particularly common structure through which turn-taking is managed in pedagogical interactions. IRE stands for Initiation-Response-Evaluation (Mehan 1979a), which is similar to the Initiation-Response-Feedback (IRF) sequence identified by Sinclair and Coulthard (1975). The teacher initiates the sequence (I), usually with a question, the student responds (R), and then the turn returns to the teacher, who often evaluates (E) or gives feedback (F) on the students' response. An example is given in Extract 3-10 where in turn 9 Tina initiates the sequence by asking Seth to describe the meaning of 'biased', which Seth then describes in turn 10, and Tina evaluates positively twice in turn 12 before repeating and rephrasing Seth's response.

The IRE sequence dominates a great deal of classroom interaction (Hogan, Rahim, Chan, Kwek, & Towndrow 2012; Kyriacou & Issitt 2008) but despite the discussions around its use in the literature, it is in itself neither a good nor a bad structure. In the literature it can be portrayed as an aspect of classroom interaction that limits students' participation or learning (Alexander 2005; Cazden 2001; Franke et al. 2007). The structure has also been used to explain why teachers often do the majority of the talking in lessons (Walsh 2011). Its advantages and disadvantages will depend upon how it used by both teachers

```
9    Tina:    lovely. can we describe it without using the
              word fair? you're completely correct, but can
              we describe what biased is without using the
              word fair. go on Seth
10   Seth:    its like equal to each time
11            (.)
12   Tina:    lovely, so not equal chance. lovely. so bias
              is when there's not an equal chance of
              getting all the outcomes. can you give me an
              example of something that would be biased
              Susy?
```

Extract 3-10 An example of an IRE sequence.

and students (Ingram, Andrews, & Pitt 2017; Roth & Gardner 2012; Wells 1993).

There is a wealth of research, usually using discourse analysis rather than conversation analysis, examining the range of ways in which teachers use the I and the E moves in mathematics classrooms, but often in isolation as individual turns rather than considering the sequence as a whole. This is most obvious if you look at the research into teacher questioning, or the use of the initiation move. Smith and Higgins (2006), for example, argue that it is the nature and quality of the feedback given in the third move that supports an interactive learning environment. However, this approach ignores the reflexive and contingent nature of the sequence (Lee 2007, 181), and risks labelling turns based solely on their grammatical structure and content, and not on how the students respond to them.

The IRE sequence is also not a structure that is specific to just classrooms, as it appears in other pedagogic interactions such as mother–child conversations (Drew 1981), which infers that it a structure relevant to learning interactions (Seedhouse 2004). Conversation Analysis handles this IRE sequence in a subtly different way and offers both descriptions and explanations for the domination of this pattern of interaction in classrooms. Within CA, this structure is a question–answer adjacency pair followed by a *sequence closing third* which is specifically designed to end the talk within that sequence (Schegloff 2007). As Wong and Waring (2009) showed, evaluations such as 'very good' are usually followed by the teacher moving on to the next question. Conversation Analysis also enables a detailed description of how this sequence offers opportunities or constrains different forms of participation depending on how it is used, which has consequences on the opportunities for learning (Waring 2009). This structure or pattern of interaction can also be considered a classroom norm, as both teachers and students orient to this structure when they interact. This IRE structure can also be embedded within a broader sequence of IRE structures which build on each other (Hogan et al. 2012).

Teachers' questions are frequently 'known-answer' questions (Mehan 1979b), in that the teacher themselves knows the answer. Yet if the student's answer is what the teacher expected, the answer itself is not evidence of learning, as the student was able to give the answer. These teacher questions can therefore often be considered as indirect requests for information rather than genuine questions; it is not the answer to the question that the teacher is interested in, but the information of whether the students know the answer to the question. However, as I show in Chapter 5, this is not the only role of these types of question. They also serve to make the information the teacher is asking about public in an interactive way, which avoids the traditional 'lecture'. As Macbeth (2001) states, the use of these known-answer questions 'organizes the room with the assurance that knowledge is already in place, and thus organizes classroom instruction as a process of revealing it' (p. 37). This reflects a norm of eliciting over telling (Macbeth 2011). The difference in use is revealed by the way that the student response is treated by the teacher. For example, both correct and incorrect answers, as well as inappropriate answers (answers that do not match in type), all give the teacher information as to whether the student(s) knows the answer. Consequently, any response which gives the teacher this information that they are looking for would be a preferred response if the teacher is using the question to assess what students know.

In contrast, non-answers and inappropriate, or disaligned, responses do not support the information being made public and are hence treated by the teacher as dispreferred in this context. It is the teacher's evaluation turn that gives us information about what the teacher was doing in their initiation turn. This context-dependent use of the IRE sequence led to Greenleaf and Freedman's (1993) approach of defining preference in terms of whether the teacher makes use of the student's response in the interaction that follows. The preference organization of an IRE sequence depends on the actions involved within the sequence.

One further thing that is evident in the data is that students mark their responses to teacher initiations far more often than is found in question-answer conversation patterns elsewhere. This could indicate that the students are treating their responses as a dispreferred response, one that differs from what the teacher is anticipating. The IRE sequence dominates most classroom interactions and students orient to this by anticipating the third turn, which frequently includes some form of evaluation of their response. By hedging their response or hesitating, and consequently marking it as dispreferred, they are also mitigating any possible negative evaluation that follows in the teacher's third turn (Rowland 1995). This also shows how students orient to the institutional role of the teacher as someone who makes assessments or evaluations. There are several possible reasons for why a student might treat their response as dispreferred. They may not be able to distinguish for themselves whether their answer will be a preferred or dispreferred responses, which may or may not relate to their knowledge or understanding of the topic. They could be treating their response as dispreferred as they are not sure it is correct, but there are also other reasons why a teacher might treat their answer as dispreferred that do not relate to whether the response is correct or not. The audience of this response is also the class as a whole, so a student may not be treating their response as dispreferred, but rather they are avoiding giving the response (Bilmes 1988).

This has implications for the mathematics that students are experiencing in their whole-class interactions. The ability to check your own answers is an important part of working mathematically, and students also need to develop their awareness of the reasonableness and appropriateness of their answers before they check them. The prevalence of the IRE sequence in classroom interactions, combined with the observation that students frequently mark their answers in some way, means that in these interactions the responsibility for making assessments and evaluations is treated as being the role of the teacher.

Looking specifically at hesitations and hesitation markers at the start of a student's response points to another aspect of the structure of turn-taking in classroom interactions. Whole-class interactions often begin with the teacher asking the class a question as a whole class: a one-to-many situation. At the end of the teacher's initiation or question, the teacher usually nominates which student can respond, often by name, but also using other nomination techniques such as gesturing. It is not usually until the end of the teacher's initiation that the nominated student knows that the next turn is theirs. Using hesitation markers at the start of their response does two things. Firstly, it avoids the dispreferred response of giving no response at all (silence) by indicating that they have accepted the next turn as theirs and an answer might be forthcoming (Wooffitt 2005). Secondly, it gives the student some time to think about and formulate their response. In many classrooms when students self-select to respond to a teacher's question, or they have bid for the turn by raising their hand, or when the interaction is between a teacher and a particular student rather than the whole class, these hesitation markers are far less common.

This treatment of responses as dispreferred may also relate to social aspects of the interaction, rather than the mathematical content of the response. A student may not want to appear confident or sure to their peers and may be marking their response as dispreferred as a face-preserving move. They could also be avoiding giving the response quickly in order to not appear too keen. This social aspect of classroom questioning raises some questions about the meaning of preference within classroom interactions. Whilst Schegloff (2007) and Pomerantz (1984) use the prevalence of hesitations, hedging, and accounts to 'define' a dispreferred response, Bilmes (1988) describes the hesitations as 'reluctance markers' (p. 173) and argues that whilst these markers are usually associated with dispreferred responses, they can also mark preferred responses and consequently cannot be used to define a dispreferred response. Instead of marking the following response as dispreferred, they are instead marking the student's reluctance to give the response, possibly because of how it may make them appear to their peers. Bilmes argues that the frequency of marked dispreferred responses only tells us that students frequently mark their dispreferred responses, and these markings do not mean that the students themselves infer a dispreferred response.

There has also been a great deal of focus on the evaluation turn of this interactional sequence. From a CA perspective, this third turn evaluates the appropriateness of the response to the initiation, which may or may not include an evaluation of the correctness of the content. It also demonstrates a teacher's comprehension of the student's response (Schegloff 2007). Mehan (1979a) focused on the methods used by teachers when the student's response did not match what the teacher expected. When teachers repeat or rephrase the question after the student's response, Mehan showed how students interpreted this as indicating that their response was inappropriate or incorrect. In contrast, when a teacher moves on to a new question, students treat this as implying that their response was correct and appropriate. Neither of these teacher turns needs to include an explicit assessment or evaluation for the students to infer whether their response was appropriate or not.

A common feature of the third turn is the teacher repeating or rephrasing the response that the student has given. Whilst Brophy and Good (1986) argue that this 'wastes time, lessens the value of pupil responses, and fails to hold students accountable for attending to what their classmates say' (p. 353), this restating or rephrasing can introduce and reinforce the use of mathematics-specific vocabulary or ways of saying things, including modelling the use of technical language within a complete sentence. It can also direct students' attention to pertinent parts of the responses that can be built on in later teaching (Ingram & Andrews 2019). Cazden (2001) also suggests that rephrasing student responses can enable a focus on the structure of what is being said, rather than the meaning.

The complexity of the social actions performed in a repetition or formulation (Heritage & Watson 1979) of students' responses has received considerable attention within CA approaches to classroom discourse. These formulations can hold 'a moment up for inspection' (Barnes 2007), enabling students to see what they've said. Yet within the repetition or formulations teachers can preserve, remove, and transform features of a student's response.

Furtak and Shavelson (2009) use the phrase 'reconstructive paraphrase' when teachers reformulate students' turns to be more acceptable or to include preferred terminology, even when this formulation changes the meaning of what the student said. Kapellidi (2015) also distinguishes between two types of formulations—those that belong to the teacher's epistemic domain, and those that do not. These transformations of students' responses can involve generalising, abstracting, or exemplifying what the student has said, or can involve initiating or performing a repair. Formulations involve some interpretation of the students' response by the teacher (Solem & Skovholt 2019) and demonstrate the teacher's understanding of what the student said. Selecting and supplementing what students said also enables teachers to connect to or develop a coherent topic trajectory, in line with the pedagogical goal of the interaction. Within mathematics education, revoicing is often the term used to describe those formulations that invite confirmation or disconfirmation from the student whose response is being repeated.

Revoicing (O'Connor & Michaels 1993) describes a particular type of formulation or way in which a teacher can use the third turn. It describes how a teacher paraphrases, clarifies, or comments on the relevance or importance of a student's response in a way that leaves the student at the centre of the discussion, whilst also identifying the response as appropriate. It is a way of 'animating' the student's response, giving it a status of something worth exploring or discussing (Goffman 1981). Structurally, when the teacher has revoiced a student's response, the turn returns to the student to affirm, or disagree with, the teacher's revoicing. In Extract 3-11 Scott is talking about equilibrium, and Tess summarizes a key idea from Scott's turn, beginning her turn with 'so', which shows that she is following on from Scott's response, but the next turn is taken again by Scott to affirm Tess's summary.

Importantly, this sequence does not follow the usual IRE structure, and the teacher is not evaluating Scott's turn in turn 85 but is using it and allowing Scott to evaluate whether her interpretation of what Scott has said is

```
84  Scott:    but equilibrium kind of isn't just about
                moving, because you know if something doesn't
                have a resultant force it continues at the
                same speed  from F equals M A. but also if
                you have like a steering wheel and it's
                constantly turning, isn't that technically a
                translation of equilibrium as well?
85  Tess:     so then we're saying equilibrium is not just
                still,
86  Scott:    yeah.
```

Extract 3-11 Revoicing where student affirms the revoicing.

acceptable. Forman and colleagues found that where teachers used revoicing, students were more likely to 'initiate explanations, provide answers or claims backed by explanatory grounds, warrants and backings, and to evaluate their own and each other's arguments' (Forman, Larreamendy-Joerns, Stein, & Brown 1998, 546). Revoicing not only affects students' positioning in relation to the other participants, it also affects their positioning in relation to the mathematics. In considering the structure of the interaction, revoicing is structurally very different from teachers repeating, echoing, or reformulating students' responses, which all act to evaluate the responses in relation to the pedagogic aims of the interaction.

Each of the analyses above of the third turn has taken a functional approach or categorization approach (Nassaji & Wells 2000). In the third turn teachers are not only evaluating or giving feedback on the 'correctness' of the pupil's turn, but are also responding to how this second turn is produced (Lee 2007). Attempts to categorize the third turn into echoes, evaluations, reformulations, etc. are unable to capture the complexity of this move and the relationship it may have with the learning of mathematics. Each part of the IRE triad is reflexively related to the others, and thus from a CA perspective the analysis of any one of these parts must consider this reflexive relationship with the other two parts. This IRE structure also occurs when it would be difficult or inappropriate to evaluate the 'correctness' of a student's response. Here, taking a CA perspective, this third move deals with how the response turn enables the interaction to continue smoothly; that is, whether the response contributes to the topic of the interaction.

Funnelling

In this section I now focus on a commonly found sequence of IRE sequences called funnelling. The funnelling pattern is initially described by Bauersfeld (1980, 1988) as a series of teacher questions and student responses that have particular features. Firstly, the sequence follows an incorrect answer from the student, or some other form of difficulty with the mathematics. The teacher then uses 'more precise, that is, narrower, questions' (Bauersfeld 1988, 36) to 'lead' the student to the correct answer. This narrowing effect of questions towards a particular correct answer (hence the term funnelling) contrasts with sequences of questions that leads students step-by-step through a process (e.g. Herbel-Eisenmann 2000); however, in both situations students do little more than complete the teacher's sentences and follow a path or reasoning controlled by the teacher (e.g. Franke et al. 2009). The literature is now awash with various examples which have led to the introduction of further terms, such as leading questions (e.g. Franke et al. 2009), guiding questions (e.g. Moyer & Milewicz 2002), and scaffolding (originally defined in D. Wood, Bruner, & Ross 1976), becoming associated or even used synonymously with funnelling.

The distinction between these terms and the precise relationship with funnelling is often not made, and consequently funnelling has become used more broadly in the literature to describe any sequences of IRE patterns that lead students through a series of specific narrow questions, often only requiring short factual responses from the students. Yet there is a difference in what is achieved through a series of IRE sequences that serve to make key facts and relationships public for later work, for example (Temple & Doerr 2012), and a series of related IRE sequences that 'lead' a student to a particular answer. In Extract 3-12 which is a simplified version of Extract 4-7, the students have been discussing the meaning of some words associated with probability that Tina has written on the whiteboard. The interaction is later followed by an activity where students are individually tossing a coin twenty times and then combining the results as a whole class. In the interaction itself, no connection is made to the tasks that came before it or after it. In the extract, Tina asks a

```
103 Tina:     lovely. so what would be the probability of
              rolling an e:ven (.) number? how many e:ven
              numbers are there? Sam (0.4) how many even
              numbers are there (0.3) o:n a: dice?
104           (1.1)
105 Sam:      u:m ↑three
106           (0.6)
107 Tina:     what are the total (0.4) numbers on a dice?
108           (0.5)
109 Sasha:    six?
110           (0.8)
111 Tina:     okay so the probability (0.6) of rolling an
              e:ven number would be what,
112 Susie:    three sixths.
113           (0.9)
114 Tina:     pardon
115 Susie:    three (.) or [a half]
116 Sid:                   [half. ] a ha[lf ]
111 Tina:                               [thr]ee sixths.
              which you're quite right, is a half. okay are
              we happy with that? are we happy with how to
              find the probability of an event?
112 Ss:       yeah
113 Tina:     beautiful. okay. …
```

Extract 3-12 Sequence of questions that make knowledge public.

series of questions that lead the students through a step-by-step process, only requiring the students to give short factual answers.

Although Steve's responses are given hesitantly, as marked by the pauses, ums, and phrasing the response as a question, Tina does not orient to these as being an issue of uncertainty. Tina's questions and her acceptance of the answer 'three sixths', rather than a half, which is only acknowledged, focuses the interaction on the identification of the numerator and the denominator and then using these to give the probability of rolling an even number. The interaction ends with an understanding check before moving on to the next task. There is little in the interaction that reveals the extent to which the students as a whole could calculate the probability themselves. This is one of the features of the funnelling pattern that Wood (1994) draws attention to: that it can give the impression of learning even though it is the teacher that has done the cognitive work. However, what the teacher has done through this interaction is to explicitly make the process public and involve the students in this process (as opposed to just telling them how to calculate the probability). Subsequently, the ability to calculate the probability of an event is treated as shared in the following task, where the students have to calculate the relative frequency of getting a head when tossing a coin. So, whilst the students are only responding to the immediate initiations, the funnelling pattern of interaction does make public knowledge that is needed later in the lesson.

This purpose of making (assumed) shared knowledge public is demonstrated further in other examples where incorrect responses are ignored, such as the second student's suggestion of 'larger' in turn 40 of Extract 3-13. This extract comes from a lesson focused on solving linear equations, where the majority of the lesson is spent with students working independently through a set of differentiated exercises. Just before the interaction in this extract, a student asks a question about the difference between an expression and an equation.

Again, the teacher leads students through a series of closed questions requiring short factual responses. The questions that Toby asks offer students the opportunity to talk about newly learnt concepts and new terminology, such as simplifying and collecting like terms, in a similar way to the example offered by Temple and Doerr (2012). Each use of a mathematical term is connected to the specific example, $3x + x$ becoming $4x$, and $3x + x = 4x$ being an expression is contrasted with $4x = 12$ being an equation. Toby ignores student responses that do not align with the use of the language he is focusing on. The series of IRE sequences are providing students with opportunities to use mathematical terminology that they are becoming familiar with, and hear it

```
35  Toby:    …wha- I mean what process, what mathematical
             process essentially has S1 demonstrated here
             by doing three ex and x
36  S1:      addition.
37  Toby:    addition. and in which case what's happened
             to that expression?
             (1.0)
38  S1:      i[t's been answered]
39  Toby:     [it's become,    ]
             (0.5)
40  S2:      larger
             (1.5)
41  S3:      um
42  S4:      it wha[t     ]
43  S5:           [simpli]fied.
44  Toby:    simplified. absolutely S5. you've got three
             ex plus x, (0.8) we've si:mplified that. by
             collecting our,
             (0.9)
45  S:       like [terms]
46  S:           [terms]
47  Toby:    like terms, good. and we've managed to say
             three ex plus x has become fou:r ex. so
             that's simplified. But it's still an
             expression. (1.2) what's- how can I turn this
             fou:r ex here … how can I turn that four ex
             from an expression (.) into an equation
48  S:       make ex three.
             (0.7)
49  Toby:    right, so if I make ex three, four times
             three would be twelve. that's still an
             expression.
             (2.0)
50  S:       give er: would it be give what ex i[s ]
51  S:                                          [oh] take
             away the e[x and ((inaudible))]
52  Toby:              [so (.) go on then Simon] hang on
53  Simon:   would it u:m you would do ex equals, you
             would- can't remember
54  Toby:    so (0.6) if I did four ex equals: (2.8)
             twelve?
55  Simon:   yeah
             (0.9)
56  Toby:    that's no:w (.) an equation. …
```

Extract 3-13 The difference between an expression and an equation.

used in a mathematical way by the teacher. If the support Toby is giving through his questioning and phrasing of his responses is withdrawn over time until the students are using the language in their own descriptions of their work on mathematical tasks, this series of questions could be considered a form of scaffolding (D. Wood et al. 1976).

Margutti and Drew (2014) have shown that within these sequences of 'funnelling' questions the teacher's assessment of the students' responses follows a pattern, with teachers using a simple repetition of a positive assessment. In contrast, in other interactions, including a series of IREs where a line or argument or reasoning is negotiated, the teachers' evaluations were formatted differently in each of the 'E' moves.

Brousseau (1997) argues that a sequence of 'funnelling' questions disguises the mathematical knowledge that is being targeted by the interaction as a whole, which he describes as the Topaze effect. And although the 'atomisation' of mathematics is currently popular in many classrooms, such interactions or tasks mean that students do not need to think about mathematical relationships, patterns, or structures in order to answer the teachers' questions (T. Wood 1998) or solve the tasks. As Wood (1998) argues, funnelling patterns of interaction only oblige students to respond to the surface linguistic patterns in order to respond appropriately to the teacher's initiations. That is, students only need to recognize the interactional structure and conditional relevance of the response they need to give following the teacher's initiation. However, it is not always the case that the mathematical knowledge is being disguised. A pattern in questioning can also be a means to make a mathematical structure explicit. Series of questions can make assumed knowledge publicly available for subsequent work, offer opportunities to use mathematical language, and perceive patterns in mathematical processes. This is just one pattern of interaction commonly found in many mathematics classrooms that contributes to students' experiences of mathematics.

Wait Time and Pausing

The turn-taking structure of whole-class interactions or teacher–student interactions allows for pauses between turns. If it is the teacher who is speaking, then they can nominate the next speaker or continue to talk. There is no possibility for any student to speak without it being sanctionable. This allows the teacher to pause. This pause can be during the teacher's turn, for example when they are explaining something, or can be after the teacher has asked a question but before they nominate a student to speak. If it is a student who is speaking, then if they do not nominate a next speaker, the next turn is the teacher's, and again the teacher can pause before speaking. Although if the teacher does not take this next turn, other students can, these students will need to leave time to ensure that the teacher is not going to take the turn (Ingram & Elliott 2014).

Similarly, students can pause at the beginning of their turns once they have been nominated. The nomination secures the turn as theirs. In ordinary conversation, pausing like this usually indicates some sort of trouble of difficulty,

```
8    Todd:      …what does that (.) produce. ((clears
                throat)). what does that produce Sonia?
9    Sonia:     um: (2.6) three hundred and five over two
                hundred and fifty
```

Extract 3-14 Students can pause at the beginning of their turns.

but in the classroom students can pause for far longer before the teacher reacts as if there is a problem. In Extract 3-14 Sonia demonstrates that she knows the turn is hers by starting her turn with 'um'; she then pauses for 2.6 seconds without being interrupted, before answering the question Todd has asked.

These pauses are structurally built into the rules of turn-taking in formal classroom interactions. Rowe (1972) suggests that there appears to be a threshold or default maximum length of pause of around 1 second that both teachers and students will allow before speaking in interactions. Jefferson's (1989) analysis of ordinary conversation also showed a 'standard maximum silence' of around 1 second before participants began to treat the silence as an indication of trouble. It would seem that whilst the structure of turn-taking in formal classroom interaction does allow for longer pauses, it may be the interpretation of longer pauses in ordinary conversation as sources of trouble that teachers and students are orienting to in their interactions (Ingram & Elliott 2016).

There are occasions where there are pauses longer than 1 second between teachers and students talking, and within a student's turn. Many of these are treated by the teacher as indicating that there is a problem. In Extract 3-15, for example, Tyler's question posed in turn 77 is hesitantly answered by Salma. There is then a short pause of 0.6 seconds before Tyler rephrases Salma's response as a question, suggesting there is a problem with this answer. This is followed by a longer pause of 1 second where Salma acknowledges that the answer is not a half. No attempt is made here to offer a different solution or to indicate that Salma understands why a half is not correct, and Tyler does not offer Salma the interactional space to do this, taking the next turn without leaving another pause. Tyler confirms that the fraction shaded is not a half before redirecting the question to another student, Sean. Sean gives another answer of 'a quarter'. Again this is followed by a long pause, this time of 1.4 seconds, before Tyler explains why the answer of 'a quarter' is not what he is looking for (though a quarter is relevant to the image on the whiteboard as a response).

In Extract 3-16 Todd has asked his students what they understand by the word 'proof'. Steve offers his thoughts in turn 16. There is a long pause of 1.6 seconds before Todd partially repeats Steve's response. This partial repeat

```
54  Tyler:    but ↑what fractio::n (0.8) what >fraction of
                that< triangle have I shaded (0.8) what
                fraction of that triangle have I actually
                sha::ded (.) Salma,
55  Salma:    um (.) w- ↑half?
56            (0.6)
57  Tyler:    >have I shaded a ↑ha::lf
58            (1.0)
59  Salma:    no
60  Tyler:    I haven't shaded a ha::::::lf. (1.3)
                S:::[ean    ]
61  Sean:         [a quar]ter
62            (1.4)
64  Tyler:    I'm shading a quarter each ti::me but I'm
                shading a quarter of a quarter of a quarter
                of a quarter so (.) it's not going to be a
                quarter exactly (.) look at it look at it in
                ↑ro::ws
```

Extract 3-15 Tyler responds to pauses as being an indication of a problem.

```
15  Todd:     right so you have (.) a reason for believing
                it or a reason that could convince somebody
                else, yes. um Steve?
16  Steve:    you can fake um proof um (0.6) about things
                (.) but you can't fake numbers.
17            (1.6)
18  Todd:     you can fake proof
19  Steve:    yes
20  Todd:     right, what yo- what are you thinking of=
```

Extract 3-16 Todd responds to pauses as being an indication of a problem.

initiates a repair and indicates the trouble sources by only repeating the part of Steve's answer that is problematic. However, in this interaction Steve does not treat Todd's turn as indicating that there is a problem, and confirms his answer. This may be because Todd has asked for his students' understanding of the word 'proof', rather than for the meaning of or definition of the word 'proof'. Therefore, Steve has given an appropriate answer, as it is part of his understanding of the word 'proof'. Todd, on the other hand, indicates a source of trouble in Steve's response—potentially that this response is not consistent with the mathematical use of the word 'proof'.

In both Extract 3-15 and Extract 3-16 the teacher has paused before indicating in their turn that there is a problem with the students' answers. These longer pauses are only present when there is a source of trouble in the interaction, and they disrupt the smooth progression of the interaction.

The end of a teacher asking a question gives rise to a TRP, and the question makes an answer conditionally relevant in the next turn. The students will be aware of this, but until one of them has been nominated to answer, they do not have the right to take the turn. Instead, in many classrooms the students raise their hands to bid for the turn.

```
54  Tyler:    but ↑what fractio::n (0.8) what >fraction of
                that< triangle have I shaded (0.8) what
                fraction of that triangle have I actually
                sha::ded (.) Salma?
```

Extract 3-17 Teacher repeats question before nominating a student.

In Extract 3-17 Tyler asks the question three times, with two pauses of 0.8 seconds between reformulation of the question. During these pauses no student speaks, no student self-selects to take the next turn, and an answer is not given until the teacher has nominated a specific student.

Rowe's (1972) and Tobin's (1986) research into the effects of teachers leaving pauses (which they call wait time) during their questioning identifies several student outcomes, many of which can be explained by the structure of turn-taking and the preference organization of repair in classroom interaction. For example, when teachers wait longer after a student has given a response, their responses increase in length and they are more likely to include justifications or support for their explanations. A pause following a student's initial response, where the teacher does not speak and the other students do not have the right to speak, will usually result in the student saying more. If this pause is long enough that it exceeds the tolerance the student has for silence, they will perceive a problem with the response they have just given. This trouble could be that the teacher has not understood the response, that the response is missing something, or just that it was not the response that the teacher was expecting. Each of these would usually result in the student trying to add to or change their response to resolve the trouble. The teacher will also feel uncomfortable in this pause, and will also generally feel the need to take the next turn as it is theirs, but if they do not speak, the turn returns to the student to expand or change their response.

The number of times students failed to answer teachers' questions also decreased, and the number of appropriate responses increased, whilst the number of speculative responses decreased. Here it is the pause that follows the teacher's question that matters. Pausing after asking a question gives students time to think about their response, and time to think about how to articulate their response. Students will also feel uncomfortable as the pause exceeds their tolerance for silence, leading to them speaking to avoid the dispreferred silence.

A number of studies (Andrews, Ingram, & Pitt 2016; Black, Harrison, Lee, Marshall, & Wiliam 2003; Rowe 1986) have worked with teachers to increase wait time, both between the teacher asking the question and the student responding, and following the students' response. In all these studies, teachers found pausing uncomfortable and difficult to do, and over time the changes to the length of pauses was not sustained. Jefferson's maximum tolerance for silence (Jefferson 1989) may account for the discomfort felt by teachers. Students will also feel this discomfort, and many teachers will also want to prevent their students from feeling uncomfortable, but it is this same discomfort that results in students speaking more often and giving more detailed responses (Ingram & Elliott 2016). So, whilst the structure of turn-taking allows for longer pauses between speakers and during turns, the preference organization surrounding silences seems to continue to be a barrier to teachers increasing wait time.

Conclusion

In this chapter I have examined the structures around turn-taking and the implications these might have on teaching and learning. These structures offer opportunities but also place constraints on what teachers and students can achieve in whole-class interactions. Although some deviations from these rules are not sanctionable and result in particular actions or behaviours that teachers may be seeking, the deviation from the rules requires interactional work by the teacher to achieve these alternative structures. The most common structure of turn-taking is often described as an IRE sequence, and I have illustrated the consequences of considering this sequence from a discourse analysis approach rather than a CA approach.

Orienting to these rules of turn-taking creates an orderliness of classroom interaction that enables the teacher to maintain control of the topic. It enables a variety of students to be called upon to do different things, which a teacher can use to their, and their students', advantage. These rules also allow the teacher to control the pace of any interactions, as they control the nature of questions being asked as well as the types of responses that are acceptable. This control of the topic also allows the teacher to adapt the topic in light of how the students are responding to their initiation. It is also this structure that enables teachers to pause and give students time to think and articulate what they want to say. All this control lies with the teacher, and these opportunities can only be realized if the teacher makes use of this structure to achieve and enable the different student actions that they want, and that students need to learn.

4

Trouble in Interaction

Trouble can refer to a range of difficulties that arise in interaction. It can take the form of an error, mistake, or misconception but can also refer more broadly to any difficulty that occurs in interaction (Seedhouse 1996). This includes talking out of turn, as discussed in Chapter 3, as well as issues of not remembering or not understanding, as explored in Chapter 5. Repair is the mechanism that we use when we interact to deal with this range of troubles in such a way as to enable the interaction to continue (Chapter 2). Many teachers use the words misconception, mistake, and error interchangeably, particularly when talking about mistakes or misconceptions that are likely to arise within a particular topic. Yet many of these teachers would also make a distinction between mistakes and misconceptions if asked about the difference. This difference is, however, apparent in the way that teachers handle them when they arise in interaction, as this chapter will illustrate.

Using CA to examine how mistakes and other sources of trouble are handled in classroom interaction draws upon several constructs and tools from CA briefly described in Chapter 2. Specifically, these include adjacency pairs and preference organization. The most frequently occurring adjacency pair in classroom interaction is question–answer, including known-answer questions, though where many of these questions function more like requests. The design of a question also places constraints on the response: the preferred response needs to match in type (Schegloff 2007). For example, a question like 'what is the mean' makes an answer that defines the mean of a set of data, an answer that describes the process for calculating the mean, or an answer like '7', conditionally relevant. 'I don't know' also answers the question, even if it is not necessarily the answer the teacher wanted, in a way that 'penguin' does not. One other form of interactional trouble is where the second-pair-part (SPP) of an adjacency pair does not match the type of response that the first-pair-part (FPP) makes relevant. You would expect a question to be followed by an answer and a request to be followed by an acceptance (or denial). Other sources of interactional trouble can also include any of the marks that usually accompany a dispreferred response, such as hesitations or delays.

Patterns in Mathematics Classroom Interaction: A Conversation Analytic approach. Jenni Ingram, Oxford University Press (2021). © Jenni Ingram. DOI: 10.1093/oso/9780198869313.003.0004

Repair in Classroom Interactions

The preference for self-repair over other-repair in ordinary conversations is also the case in classrooms. However, classrooms interactions involve multiple speakers, and consequently there are more options for who can initiate and who can perform a repair on trouble in interactions. The different turn-taking structures between classrooms and conversations described in Chapter 3 also influence who can initiate and perform repairs. The preference for self-repair in ordinary conversation is most identifiable from the prevalence of this type of repair, but it is the structure of the repair that makes it preferred rather than the prevalence. This is evident in classroom interactions, where self-repairs are far rarer but are still treated as preferred. One way in which this happens is by the teacher initiating the repair, but not performing it. In Extract 4-1 Trish locates the problem with Sonia's answer, but it is Sonia who gives the correction in turn 279. Trish has initiated the repair, but it is Sonia who performs it.

By initiating the repair Trish has indicated that there is some source of trouble in what Sonia has said, but she has also precisely located what the source of trouble is, whilst still allowing Sonia the opportunity to perform the self-repair. Teachers also do interactional work to enable a peer-repair, thus avoiding a teacher-repair, in a similar way to the way Trish enabled a self-repair in Extract 4-1. In Extract 4-2 Trish negatively evaluates Simon's answer but indicates that it has matched in type and the answer is close to the answer she is looking for. It is then Steve in the next turn who gives the answer that Trish is looking for. The negative evaluation is constructed in such a way that it locates the source of the trouble but it does not perform the repair. This extract is somewhat unusual in that Trish uses a negative evaluation to locate the trouble. Negative evaluations are generally rare in whole-class interactions, but, as I discuss below, there are interactional contexts where they enable the interaction to return to its smooth trajectory as soon as possible, as is the case here.

```
276 Trish:    why, where are you getting those numbers from
              though
277 Sonia:    because it's (.) in the middle of ten
278 Trish:    ok so that is basically (.) what we're going
              to do but hold on a second. I don't agree
              with you that five is the number between one
              and te[n ]
279 Sonia:          [an]d a half
```

Extract 4-1 Teacher initiates but does not perform a repair.

```
255 Simon:    fif- no sixty
256 Trish:    no: (.) not quite
257 Steve:    nearer fifty (1.1) fifty five=
258 Trish:    =fifty five …
```

Extract 4-2 Teacher-initiated peer-repair.

```
260 Tyler:    … what is a pri:me number. Sibyl.
261 Sibyl:    an number that can only be like divided by
              itself.
262 Shaun:    and one
263 Sibyl:    and one
264 Tyler:    and one. good. ok
```

Extract 4-3 A student initiates and performs the repair.

On the other hand, when other students initiate a repair on one of their peer's turns, they often also perform the repair in the same turn and do not offer the original student the opportunity to self-repair. In Extract 4-3, Shaun both initiates and performs the repair on Sibyl's turn by adding in the missing information. Sibyl accepts this repair by repeating it before Tyler positively evaluates it.

Sibyl's first answer is appropriate, but incomplete, and so Shaun's repair is not an explicit negative evaluation of Sibyl's answer but builds on it, as indicated by Sibyl's acceptance of Shaun's turn and Tyler's positive evaluation that closes down the interaction. Each of the extracts so far in this chapter illustrates a mistake that is quickly repaired by a student.

One context where teachers treat teacher-repair as preferred over other-repair is when the trouble arises in a student's explanation. One example is offered in Extract 4-4.

Trish begins her turn by indicating that there is a problem with Sarai's turn and locates the problem in the explanations through the phrase 'it needs a little bit'. It is not clear at this point what the nature of this trouble source is. Sarai uses the phrase 'higher number' to refer to the number −40, but higher is problematic in this context. On the one hand, higher could refer to the magnitude of the numbers—40 is indeed larger than 10—but higher can also refer to a direction, and in Trish's explanation she references a thermometer and 'going up the scale'. Trish then positively evaluates part of Sarai's explanation before pausing twice, once for 1.6 seconds and then for 1.8 seconds, offering Sarai the opportunity to self-repair. Sarai does not take up this opportunity, and Trish then offers her own explanation and thus completes the repair.

```
15  Sarai:    is it minus ten because (0.6) when, when it's
              in minus numbers you take away get (.) like a
              higher number so (.) if it's minus ten then
              you take away minus thirty then it gets to
              minus forty which is the temperature range,
              (Verhoinshk)
16  Trish:    almost (0.4) almost it needs a little bit.
              you are thinking about going backwards which
              is good. (1.6) no? (1.8) the reason (.) is
              because (.) if you think about your number
              line, if you're starting off at minus forty,
              (.) if you're thinking of it as a temperature
              scale, if you're thinking of it as a
              thermometer, if you're starting at minus
              forty and its getting thirty degrees warmer,
              so you're going up thirty, you're going up
              (0.4) your scale (0.6) thirty. now if you're
              reading the temperature, if you're using the
              thermometer, it's going up the scale (0.4)
              ok. so you're adding, so you're going up the
              scale.
```

Extract 4-4 Teacher initiates and performs repair.

In this extract the nature of the trouble source is one of reasoning and understanding, as made visible through how Trish initiates and repairs the trouble. This difference is one way that troubles with knowing and remembering are treated differently from troubles with understanding or reasoning, which is explored further in the next chapter.

Enabling Students to Give a Preferred Response

Another feature of interaction that makes visible the preference for self-repair and peer-repair over teacher-repair is the interactional work teachers do to enable students to give a preferred response. In Extract 4-5, Todd and Skye are continuing an interaction where Skye has already provided an answer to an earlier related question by Todd, of which is longer, a microcentury or a lesson.

In turn 23 Todd reformulates the question asked in turn 21 to be more specific, in that he asks Skye for more information about where the numbers given in her earlier answer came from. Skye's previous answer described the process that she went through to get to a final answer, but did not include any explanation of why she took these steps. She has therefore explained how she got her answer but she has not explained why the answer or her process are appropriate (Leinhardt 2001). The trouble has arisen because Skye has already presented her final answer in the previous turn. Todd's reformulation overlaps Skye's response, resulting in them speaking simultaneously. Skye's overlapped

```
20  Todd:     so what did you do next then Skye.
21  Skye:     um m
22            [(.)  I worked it out on              ]
              a calculator
23  Todd:     [how did you get those numbers from]
24  Todd:     some people wrote that sort of thing. they
              wrote their answer it's (.) longer because I
              worked it out and it's longer. um what I
              really wanted was the details of how you
              worked it out, of what you did (0.5) um (0.3)
              who (.) can pick up the thread there. (2.2) a
              lot of people are throwing numbers around and
              I think they sort of (.) work their way round
              the class without people necessarily knowing
              where they came from.
```

Extract 4-5 Teacher reformulates the question after a student hesitation.

```
40  Todd:     and (.) I want to know what you think is the
              same or what's different about those two.
              (3.1) I mean we did all that yesterday but I
              just thought it'd be nice if we (.) stood
              back and thought about (0.6) what it meant
              and w-hat's the same and what's different
              about the right side and the left side. (3.7)
              °a hard question°. have a think. (4.1) you can
              say something that's quite obvious and that's
              fine. I'd just like people to make (0.4)
              observations about what's the same and what's
              different. (0.5) ((small group are laughing))
              what are you laughing at?
```

Extract 4-6 Many pauses during the asking of a question.

response can be interpreted as an answer to the question as originally posed in turn 20, and Todd treats it as such when he takes the next turn. Both Todd and Skye have attempted to repair the trouble: Skye by offering an answer to the question and Todd by reformulating the question to enable Skye to self-repair. Skye is, however, not offered any opportunity to respond to this reformulated question, as there is no transition relevance place following the completion of her answer of 'working it out on a calculator'.

In this next example, Todd is talking to the whole class and does not specifically nominate the next speaker, which is the norm of turn-taking in his classroom.

In Extract 4-6, Todd is referring to two visual representations drawn on the whiteboard shown in Figure 4-1, one on the left representing a numerical example and the other an algebraic representation of the relationships between the numbers given in the first diagram. Todd's question, which is asked three times, is asking the students what is the same and what is different about the two representations. There is an initial pause of 3.1 seconds, where

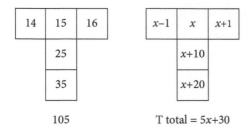

Figure 4-1 T-totals task.

```
103 Tina:    lovely. so what would be the probability of
             rolling an e:ven (.) number? how many e:ven
             numbers are there? Sam (0.4) how many even
             numbers are there (0.3) o:n a: dice?
104          (1.1)
105 Sam:     u:m ↑three
106          (0.6)
107 Tina:    what are the total (0.4) numbers on a dice?
108          (0.5)
109 Sasha:   six?
110          (0.8)
111 Tina:    okay so the probability (0.6) of rolling an
             e:ven number would be what,
112 Susie:   three sixths.
113          (0.9)
114 Tina:    pardon
115 Susie:   three (.) or [a half]
116 Sid:                 [half. ] a ha[lf ]
117 Tina:                             [thr]ee sixths.
             which you're quite right, is a half. okay …
```

Extract 4-7 Breaking down a question into its parts.

the question is phrased as a request, rather than a question. The second time Todd asks the question it is also not clear that it is a question, as he has asked the students to 'stand back and think'. After a long pause of 3.7 seconds, where no student has offered an answer, Todd offers an account for why there is no response in that it might be 'a hard question', which also serves to make it clear that his earlier statements were intended to be a question. However, this is then followed by another request to think. This is followed by another long pause of 4.1 seconds before Todd clarifies (and possibly broadens) the range of acceptable answers. As the first speaker of an FPP, Todd has 'an opportunity to change the first pair part to a form which will allow the response which is apparently "in the works" to be delivered as a preferred response, rather than a dispreferred one' (Schegloff 2007, 70). In both Extract 4-5 and Extract 4-6, the teacher has altered their original question in some way to support their students in responding, avoiding further dispreferred silences that had initially followed the original asking of the question. The structure of these alterations and reformulations is the same as that in ordinary conversation: the first speaker has to do interactional work to enable the second speaker to give a preferred response. In classrooms, this first speaker is usually the teacher.

This practice of altering a question can occur before there is any visible trouble in the interaction. In Extract 4-7 Tina begins the sequence by asking what the probability of an even number when rolling a die would be. Without pausing to offer students the opportunity to respond or raise their hands to indicate they could respond, she simplifies the question to how many even numbers there are on the die. Sam answers very hesitantly, leaving a long pause of 1.1 seconds and beginning his turn with a hesitation marker. Tina does not acknowledge Sam's response and does not evaluate it, but instead asks a new question in turn 107. This question is answered by Sasha in turn 109, again offered hesitantly with pauses and a rising intonation. This is acknowledged by Tina with an 'okay', then Tina begins her next question with 'so', indicating that it is building upon the previous answer(s) (O'Connor & Michaels 1996), and here she re-asks her original question of 'what is the probability of rolling an even number'.

Susie gives the answer of 'three sixths' in turn 112. This is followed by a pause of 0.9 seconds before Tina says 'pardon'. This indicates that there is now trouble in the interaction. 'Pardon' often indicates a problem with hearing, and Susie begins her turn by beginning to repeat her previous answer, suggesting that she has interpreted this as a problem of hearing, but then she adjusts and offers another answer of 'a half'. These two answers are mathematically equivalent, and Susie demonstrates this in her use of 'or'. Sid also self-selects to give the answer of 'a half' as well, overlapping with Susie. This inclusion of a half suggests that both Susie and Sid have interpreted Tina's pardon as indicating a problem with the original answer of 'three sixths'. However, Tina accepts the answer of 'three sixths' in turn 117, showing that the original source of the trouble was one of hearing. She also accepts the answer of 'a half'.

In contrast to the two previous examples, Tina's reformulation of the question has simplified the mathematics that the students have to do, and this is an example of what Bauersfeld (1980) refers to as a 'funnelling' pattern, as discussed in Chapter 3. There is nothing in the interaction to indicate whether the students could or could not have answered the original question without the simplifications made by Tina. What has been done in this interaction is that both the answer and the process of attaining the answer have been made explicit and public to the whole class. This can then be taken as shared in the interactions that follow and is in fact needed for the task that Tina asks her students to work on independently later in the lesson (Ingram, Andrews, & Pitt 2017). This funnelling is another example of an interactional pattern that is in itself neither good nor bad; its effect depends upon its use and purpose. Teachers can also support students in giving a preferred response by referencing and making public any assumed shared knowledge, or by making references to occasions where they worked together on a particular piece of mathematics, in their talk preceding the asking of the question.

One advantage of whole-class interactions is that teachers have a range of students they can nominate to answer a question. The turn-taking structure outlined in Chapter 3 means that students can only speak if the teacher nominates them. Teachers can thus ask a question and nominate who should answer after students have bid for the turn in some way, whether by raising their hand or by using their body language to make themselves available (Käntä 2012). This increases the probability that the response given will be the one the teacher is looking for. Another strategy would be to allow students to interact with each other to construct a response. In Extract 4-8 Tyler asks what the meaning of direct proportion is. There are two long pauses during his turn, during which students could bid to take the next turn or offer a response. During this time, only one student raises their hand to bid for the next turn. Tyler then asks the students to talk to each other before responding to the question, more commonly known as think-pair-share (Lawler 2018; Schoenfeld 2018).

Offering students the opportunity to talk through what they think the meaning of direct proportion is, is likely to increase both the number of students able to answer the question and the number of students willing to do so. This also offers the teacher the opportunity to 'overhear' what the students are talking above and thus identify a student who might be able to give a preferred response.

Students also do interactional work to avoid giving a dispreferred response. They often mark their answers in some way and often give them hesitantly

```
24  Tyler:     ... ok what do I mean by that, what do I
                mean, if I said to you two quantities are in
                direct proportion what do I mean by that.
                (2.0) if I said two things are in direct
                proportion, what do I mean. have a guess.
                (3.1) ok in fact. (0.7) talk to the people on
                your table first
```

Extract 4-8 Teacher invites students to talk to each other first.

and using a questioning intonation, as they did in Extracts 4-4 and 4-5. This can mitigate against a negative evaluation or initiation of trouble by the teacher in the turn that follows. It also obliges the teacher to make an evaluation, as the hedging indicates uncertainty in the student's answer. However, it also relates to the nature of the FPP, where teachers are often asking questions they already know the answer to ('known-answer questions'). The frequency of preference markers in student responses is considerably lower when the students are reporting on what they have done or are offering their opinions or thoughts, i.e. in situations where the response is outside the teacher's epistemic domain and it is the student who decides what is correct or not, or where this evaluation has occurred elsewhere, such as in group work or one-to-one interactions with the teacher earlier in the lesson.

Avoiding Dispreferred Evaluations

In this section I look at what happens in an interaction where the responses given by students are treated as an incorrect or inappropriate response to a teacher's question; that is, where the trouble source is in the content of a student's answer. In most cases an incorrect answer is a dispreferred response to a teacher's question, but there are contexts where this might not be the case. If the teacher uses and builds on the incorrect response given in a way that enables the interaction to continue smoothly, then this incorrect response is treated by the teacher as preferred. This might arise, for example, in the case where the teacher has asked a question where a common misconception is likely to be offered in the answer, which they can then use to explore why this answer is not correct.

Negative evaluations of students' responses are very rare in classrooms (Ingram, Pitt, & Baldry 2015), and teachers do a lot of interactional work to avoid making negative evaluations whilst still indicating that there is trouble in the student's response. Extract 4-9 offers an example of this. Sam has given an answer to a question on a worksheet as ten metres. Todd repeats this answer before offering a long pause, which gives Sam an opportunity to affirm or change his answer. Sam does not say anything, and Todd continues by asking the other students if they want to agree or offer a different answer to Sam. The other students are offered the opportunity to volunteer to offer another response through Todd's pauses of 0.8 and 0.9 seconds during the question. Todd also does not emphasize the words 'same' or 'different', so is not giving any clue as to which he prefers. He then repeats Sam's answer in full, including the previous turns in the interaction not included here.

There is also a contrast in the use of pronouns here, with the 'point one seconds' as something we have, which is the number given in the question, and the rest being attributed to Sam and using the words that Sam himself used. Yet this is not an example of revoicing, as Sam is not directly invited to confirm or deny Todd's rephrasing of his answer. Following this, there is a long pause of 2.3 seconds which offers all the students the opportunity to bid for the turn, and now also offers Sam an opportunity to add to or alter his answer. No student takes up this opportunity. Todd then reports that he knows that other students have a different answer, before Sophie bids for the turn and Todd invites her to take the next turn.

At no point during Extract 4-9 does Todd negatively evaluate Sam's answer. In contrast, following Sophie's different answer Todd repeatedly positively evaluates this new answer in a range of ways in turn 47. The pauses during Todd's turn 45 indicate that there is a source of trouble in the interaction, but Todd is not explicit about whether that trouble is with Sam's answer or with the other students not answering the prompt 'what do other people think about that'. Todd has not located the error.

Another interactional context that enables a teacher to avoid giving an evaluation is where they do not select which speaker to answer the question they have asked. Here we return to Extract 3-6 to analyse the structure of

```
42   Sam:      ten (0.7) because I (converted that into
               metres)
43   Todd:     ten ↓what
44   Sam:      metres
45   Todd:     ten metres (1.5) what do people think about
               that. Did anyone do the same or (0.8) or
               different (0.9) so we've got ↑point one
               seconds: (.) and Sam has times'd it by a
               hundred kilometres an hour (0.5) and his
               answer (0.6) he has given is ten metres (2.3)
               >I know that not everybody did that, because
               I saw people doing different things< Sophie
               what- (0.8) what did you do (0.6) what do you
               want to=say
               ((transcript omitted))
46   Sophie:   [a]nd then we just (.) divided it by sixty
               again to make it into seconds (.) and then we
               divided it into ten, divide by ten (0.4) to
               make it to (0.3) tenths of a second for which
               we got two point seven metres
47   Todd:     ri:ght very good that's the right answer very
               good…
```

Extract 4-9 Working to avoid giving a negative evaluation.

```
268 Tyler:    o↑kay? (0.4) which is the fi::rst prime
              number.
269 Scott:    one
270 Sean:     tw[o
271 Skye:        [zero
272 Sean:     two
273           (0.4)
274 Scott:    one
275 Tyler:    one[ two
276 Sheila:      [two
              ((transcript omitted))
281 Tyler:    why on:e
282 S?:       one isn't (?)
283 Tyler:    ↑why is it not a prime number
284 Sean:     >cos you can only divide it by,< (0.8) one,
285 Scott:    yeah but you ↑can ↑divide it ↑by itself, cos
              it divides by itself
287 Sheila:   yeah but you ↑can't (.) you can only divide
              it by one thou:gh
289 Scott:    ↑yeah and that's dividing by it↑self,
290 Sheila:   yea::h but (.)↑one's (the same) numbe:r
```

Extract 4-10 Withholding an evaluation.

repair, re-presented in Extract 4-10 below. The self-selection of speakers here results in more than one student speaking and offering contrasting answers. In turn 275, Tyler repeats two of the answers but offers no evaluation or further indication of which one is the answer he is looking for. The interaction continues, with different students continuing to offer the answers of one or two, until in turn 281 Tyler asks another question. This question asks why one is the first prime number and then in turn 283 Tyler asks why one is not the first prime number. In the turns that follow, students begin to give accounts for why they are disagreeing, and it is the students that indicate that there is a problem with the answers given.

One example of a context where a negative evaluation is not avoided by a teacher is where students have given an answer that matches in type and idea but have not used the wanted vocabulary to express that idea. In Extract 4-11 Trish negatively evaluates part of Shannon's answer and immediately gives the correct terminology of '2D'.

This raises a question about why teachers avoid negative evaluations except in situations where the trouble source is the use of subject-specific vocabulary. In this situation the trouble source is in the form of what is being said, not in the content or meaning of what is being said. In some situations, the form of what is being said may be the topic or focus of the interaction; however, in all the cases where the language is corrected immediately by the teacher, it is the content of the answer that is used or built on in later interactions. Hence the immediate repair enables the topic to return to what the

```
14  Shannon:   a net is the three d shape flattened out into
                a (.) one d shape
15  Trish:     not one d but two d okay.
```

Extract 4-11 Correcting students' language.

teacher is focusing on quickly and reduces the opportunities for students' attention to shift from the content to the form. It also minimizes the disruption to the flow of the interaction (Seedhouse 2019).

This avoidance of negative evaluations gives the interactional message that errors are to be avoided, and that they are face threatening (Ingram, Pitt, & Baldry 2015; Seedhouse 1997). Furthermore, immediately correcting an error, as in Extract 4-11, enables the teacher to move the interaction on and treats the error as inconsequential to the interaction that follows. This negative evaluation and repair also reinforces the teacher's epistemic authority within the interaction. This relationship to epistemic authority is examined further in Chapter 5.

In Extract 4-9 and Extract 4-10 the teachers have not evaluated the answers initially given, but have instead provided both the student who gave the answer and other students the opportunity to repair the trouble with the answer given. This interactional work reveals the dispreferred nature of teacher-repairs, with teachers offering several opportunities for self-repair, and then peer-repair, to avoid repairing the trouble themselves. This challenges the findings in other studies where teacher-repair is more common (Liebscher & Dailey-O'Cain 2003; Macbeth 2004; McHoul 1990), though this may relate to the different contexts in which the analysis took place, such as the curriculum area and age of students, but also the norms of the classroom from which the data was collected. As there is a shift in epistemic authority depending upon who makes the repair, this difference suggests that students in the maths classrooms considered here are treated has having the epistemic access to make repairs. This then further suggests that what is being discussed is being treated as knowledge that the class has access to; that is, knowledge that is shared. It also treats errors as part of the learning process. Errors offer evidence of learning, and the preference of self-repair and then peer-repair makes this learning explicit.

Teachers can also use repair initiations deliberately to achieve a pedagogic goal. In Extract 4-12 Tess initiates a repair in turns 287, 290, 292, 294, and 296. Each of these initiations invites the students to add more precision and detail to their responses. This also opens up opportunities to discuss some of

```
285 Tess:      ...okay. so if I wanted to know (.) the moment
               er so let's say that bit can be two and this
               seven.  so how could we work out any of these
               things. (6.8) you don't get them this hard by
               the way. Stefan:
286 Stefan:    well for your (0.3) er it's going to be er
               four point five (.) times m g (0.2) minus:
               (0.5) t
287 Tess:      what are yo- what are you doing,
288 Sam:       °what are we trying to find°
289 Stefan:    we're working out (1.4) all the kind of,
               you've got to work out all the magnitudes
               ((inaudible))
290 Tess:      magnitudes of what.
291 Stefan:    well you've got to work them out (1.2)
               factoring in the distance from the (pivot)
292 Tess:      I don't understand what them is.
293 Stefan:    all the um points of forces (1.1) moments I
               suppose
294 Tess:      all the moments
295 Stefan:    yes
296 Tess:      okay and where are we taking moments about
297 Stefan:    er B because that's your pivot point
298 Tess:      does it have to be that we take moments about
               B.
299 students:  no
300 Sid:       no because, (2.1) no you can take it anywhere
               can't you
301 Tess:      we take moments about anywhere you like.
               okay. (1.7) why might it be useful to take
               moments about B. Steve
```

Extract 4-12 Deliberately initiating a repair.

the decisions the students made, such as which point to take moments about in turn 298.

Initiating a repair and performing a repair can both be used by teachers to support their students in explaining their reasoning, being precise in their explanations and descriptions, and also focusing attention on some of the implicit steps or decisions within a process. By making this structure visible through a microanalytic approach, this structure becomes available to teachers as another pedagogic tool they can use to influence the mathematics that their students are learning in interaction.

Making Use of Accounts

The preference organization of classroom interaction can also result in students offering explanations for their answers. Whilst the majority of times where a student gives an explanation is where the teacher has specifically asked for one, there are also contexts where students offer explanations or justifications without being asked to, and these share similar structural characteristics. These build on the CA idea of preference and accounts that often accompany dispreferred responses. The two situations I consider below include answering a question that has already been answered by another student and speaking out of turn.

To begin this section I return to the examples in Chapter 3 where a teacher asks a question, often undirected to a specific student, and consequently multiple students give answers. In Extract 4-13 Tyler has asked a question about proportions, though in a different context where the proportional relationships the students have been working with prior to this interaction are not relevant.

At no point in this interaction has Tyler explicitly asked for an explanation. In the interaction three different potential answers are given, one hour in turns 188, 196, and 197, two hours in turns 193 and 194, or it's a trick question in turns 190 and 192. Whilst two hours and a trick question are both answers that most maths teachers would treat as being correct or acceptable answers to the question, one hour is another response which could have been

```
182 Tyler:      … six shirts take two hours to dry on a
                washing line, how long will it take to dry
                three shirts. should be a question mark at
                the end.
183 Sarai:      what?
184 Scott:      what?
185 Sam:        the:y'd be the same wouldn't they
186 Sean:       oh that is easy
187 Sophia:     another trick que[stion]
188 Seth:                        [one ] hour
189 Tyler:      one hour
190 Sasha:      no [that's a trick [question ]
191 Seth:         [ ((inaudible))        ]
192 Simon:                          [that's a ] trick question
193 Shannon:    it'll take two hours
194 Saul:       it'll take two hours
195 Susie:      oh
196 Stefan:     (it [will take one hour.)        ]
197 Saul:           [it'll take one hour because]
198 Sid:        it will take one hour.
199 Sheila:     no it wouldn't [it would take ]two (.)
                [because (.) they've all got to dry
                ((inaudible))]
200 Shannon:                   [no it wouldn't]
201 Skye:       [((inaudible))]
202 Shannon:    [there all t-shirts aren't they]
203 Sheila:     it doesn't matter how many's on the [line.]
204 Sian:                                           [yeh ]
205 Salma:      it's all gonna take two
206 Tyler:      good. [ok? you've got to (be aware of the)]
```

Extract 4-13 Students mitigating their responses with accounts.

anticipated, as it is the answer that you get if you use the proportional rela-tionships of halving the number of shirts then halving the time taken. However, it is clear that the answer to the question cannot be both two hours and one hour. After several students have given a response to the question, some students begin to attempt to offer an explanation for their answer. In turn 197 Saul indicates that he sees an explanation as being relevant to the interaction by ending his turn with 'because', even though he does not com-plete the explanation himself. Sheila, in turn 199, does successfully give a rea-son, as does Shannon in turn 202. Both these explanations could be interpreted as supporting the correct answer of 'two hours' even though the explanations themselves do not make this explicit. The students who offer explanations are giving dispreferred responses in that they are disagreeing with answers previously given by their peers, which is a disaffiliative move. These explanations act as accounts for why they are giving a contrasting response.

Tyler often asks questions where a range of different answers are likely, and he invites these different answers by allowing his students to self-select as next speaker. We can see that this is a deliberate strategy from the occasions where the point of contention did not arise. In Extract 4-14 Tyler is asking whether the sequence of numbers will ever actually reach its limit of 5.

Tyler's question is a polar question, requiring just a yes or no response. Both the students who respond both give the same answer. Tyler leaves a pause of 0.8 seconds before speaking, allowing other students the opportunity to self-select like Susie and Stefan have before, asking why. However, no stu-dent takes this opportunity, and only the answer of 'no' is given and made public. Tyler makes no evaluation of the answers given before explicitly ask-ing for an explanation.

Disagreeing with another student's answer does not only happen when the teacher allows multiple speakers to respond to a question. In the situations discussed at the beginning of this chapter, where a teacher initiates a repair but it is a student that performs the repair, a student will often include an

```
173 Tyler:    never ending. good. never ending number,
              infinite
              number of decimal places, will you every
              actually
              hit five.
174 Susie:    no=
175 Stefan:   =no
              (0.8)
176 Tyler:    why
```

Extract 4-14 Point of contention does not arise.

```
15   Sasha:    doesn't squared mean you times a number by
               two, so twenty-four times two is forty-eight?
16   Tara:     so are you saying that twenty-four square
               centimetres means forty-eight centimetres, is
               that what you're saying?
17   Sasha:    I don't know.
18   Tara:     other people agree with that, or not?  can
               you, can you say what you think about that,
               so Sasha is suggesting that actually the
               answer is forty-eight. is that right or not?
19   Simon:    no, it depends what shape you want. if you're
               going to be doing a triangle [outside], or a
               rectangle because forty-eight centimetres,
               okay, in the area and then you will do,
               because it's ((inaudible)) a rectangle, you
               just go do that  so if only the triangle in …
```

Extract 4-15 Disagreeing with another student.

explanation for the new answer they are giving as a repair. The example given in Extract 4-15 involves a student, Sasha, asking a question that includes a mistake. Tara revoices Sasha's response and invites her to affirm this revoicing, but Sasha does not, instead saying she does not know which is consistent with her asking the question in the first place. Tara then invites other students to make the evaluation of whether Sasha's answer is right or not, and Simon does this in the next turn. He begins with a no, before giving an explanation which continues over several turns, interspersed with clarification questions from Tara, until later in the interaction, at turn 26, Simon's response is accepted by Tara.

Disagreeing by giving contrasting or contradictory answers is a dispreferred response and a disaffiliative action that needs accounting for (Heritage & Raymond 2005). The explanations offered by the students in Extract 4-13 and Extract 4-15 account for the dispreferred nature of their turns. These explanations all defend the new answer being given by the student, and do not critique or explain why the previous answer is incorrect (Ingram, Andrews, & Pitt 2019), minimizing the disaffiliative nature of their turn.

Explanations are also given by students when they do not have the right to speak due to the rules of turn-taking. There are occasions where self-selection to speak is allowed by the teacher and these are detailed in Chapter 3; however, in most whole-class interactions only a student nominated to speak by the teachers has the right to the turn. For example, in Extract 3-8, repeated here as Extract 4-16, Seth repeats Scott's answer of 'true' and adds an explanation for why it is true. This is despite the teacher nominating only Scott to give the required response, though in this case the explanation is not sufficient to mitigate against speaking out of turn.

```
235 Trish:     in we've already had j so e thank you Scott
               do you think that's true or false
236 Scott:     true
237 Seth:      true cause there's only three outcomes to get
               so we-[
238 Scott:        [shut up I was ((inaudible)) here]
239 Stefan:    shut up so it's gonna be a (third) ]shut up
240 Trish:     er excuse me. just settle down. could you not
               tell each other to shut up because it's very
               rude
```

Extract 4-16 Adding an explanation when speaking out of turn.

There are a variety of interactional contexts in which a student turn may be dispreferred. These include answering questions that have already been answered, as well as contexts where they are breaching the rules of turn-taking governing the particular interaction or are deviating from the topic being focused on by the teacher. In these contexts, adding an explanation can mitigate the dispreferred nature of the response (Ingram, Andrews, & Pitt 2019).

Conclusion

How troubles in interaction are treated by teachers depends upon the nature of the trouble itself. Both teachers and students do a lot of interactional and pedagogical work to avoid giving a dispreferred turn. However, this interactional preference organization also influences the nature of mathematics within classroom interaction.

A key feature of working mathematically is making conjectures that can be tested and rejected, modified or accepted as a result. Not only do students need to feel they can make mistakes and conjectures, however; they also need to be able to make their own judgements about accuracy or appropriateness. The dispreferred and disaffiliative treatment of negative evaluations and disagreements challenges the establishment of a conjecturing atmosphere (Mason 2016) involving respectful disagreement and argumentation. Direct negative evaluations also enable teachers to balance the tension between accurate communication and the communication of meaning.

5

Thinking, Understanding, and Knowing

Teachers of mathematics often argue that they want their students to think about mathematics, think mathematically, and to understand what they are learning. Mathematics also involves knowing a range of facts, properties, procedures, and a specialized vocabulary in order to answer questions, solve problems, and reason mathematically. Learning mathematics can also be considered as learning a range of actions or behaviours, such as specialising, generalising, and justifying. It is therefore analytically interesting to ask the question of how much these different aspects of learning mathematics are talked about and are done in mathematics classroom interactions. When considering issues of knowing or understanding, one focus of a Conversation Analysis (CA) perspective is how the words associated with cognitive processes such as know, remember, or forget are treated by teachers and students as interactional objects (Heritage 2012b). Another focus is how issues of knowing, or not knowing, etc. are treated within interactions. We do not consider what students know or understand, or do not know or understand, but instead look at how teachers and students talk about knowing and understanding, and the social actions these utterances perform as they interact. For example, when a student states that they 'don't know', this could be a request for help, or a complaint about the clarity of an explanation (Lindwall & Lymer 2011), or many other possible actions, including resisting taking part in the interaction.

Classrooms are all about knowledge and understanding, whether you prefer to talk about the transfer of knowledge, the negotiation of knowledge, or the sharing of knowledge, or, in the case of ethnomethodological perspectives, what it means to know in interaction and the actions related to knowing or understanding. As an institutional setting, there are different epistemic statuses assigned to the different people who interact in classrooms. Teachers are generally considered to be knowledgeable about the subject being taught, whereas students are generally learning this knowledge. Teachers generally ask the questions, and students generally answer them. Furthermore, teachers often ask questions that they already know the answer to (Mehan 1979b). As the earlier chapters have shown, there are interactional structures that

Patterns in Mathematics Classroom Interaction: A Conversation Analytic approach. Jenni Ingram,
Oxford University Press (2021). © Jenni Ingram. DOI: 10.1093/oso/9780198869313.003.0005

further demarcate teachers and students, many of which, within the hands of a skillful teacher, can be used to promote the primary goal of classroom interactions: learning.

The approach I take here is one that considers learning to be emerging from interaction and participation in the discursive practices of the classroom. Learning is both situated in and constituted in interaction. It is not a cognitive or individual phenomenon, and I make no claims as to what people do or do not know or understand. Instead I am interested in how knowledge, or knowing, understanding, and other cognitive phenomena, are talked about and used as teachers and students interact. This perspective also means that learning, knowledge, and understanding are not things that are acquired; rather, they are characterized by change, variability, and a connection with the context in which they are being considered.

Recent CA research has begun to examine the management of epistemics in a range of interactions. Most notable is the work of Heritage and colleagues (Heritage 2012a, 2012b, 2013; Heritage & Raymond 2005). The focus of much of this research is on how different participants orient to what they and others do or do not know, or claim to know or not know (e.g. Barwell 2013). Heritage in particular distinguishes between different positions of epistemic access, which are relevant to the interactional contexts. K+ participants are more knowledgeable and K− are less knowledgeable. Epistemic status is related to the epistemic access different participants have to the knowledge that is at stake in the classroom.

In the mathematics classroom, epistemic access is not only concerned with the knowledge that is associated with the subject matter. So, whilst teachers may be in a K+ position in relation to the mathematics being discussed, there are also cases where teachers may be in a K− position. Compare the questions asked in Extract 5-1 and Extract 5-2.

In Extract 5-1 Tristin has asked a question that he knows the answer to, and we can see this in the way he uses the third turn of the IRE sequence in turn 6, where he repeats Stefan's response and then changes topic (a way of positively assessing a student's answer (Ingram, Pitt, & Baldry 2015)). In contrast, in

```
4   Tristin:   ok that's quite good she's picked five
               numbers and they are connected, who can tell
               me what the connection is between (.) Sibyl's
               numbers. (1.3) Stefan
5   Stefan:    they're um multiples of five!
6   Tristin:   they're all multiples of five. so I'm going
               to clear those off …
```

Extract 5-1 Tristin in a K+ position.

```
26  Tanya:    … when you initially saw the question, how
                did you think of it.
27  Sian:     divide by four
28  Tanya:    who said that. divide by four,
29  Sian:     and then (divide by four again)
30  Tanya:    and then divide the answer by four and then
                a- and then what,
31  Sian:     (add them together)
32  Tanya:    okay did anyone do anything differently when
                they initially looked at it. Saul what did
                you do.
```

Extract 5-2 Tanya in a K– position.

Extract 5-2 Tanya is asking students what they thought of when they first saw the problem posed at the beginning of the lesson. Whilst Tanya might know the solution to the problem posed, and a range of different methods that students might have taken to work on the problem, she does not know what Sian initially thought of. Thus, whilst Tristin is in a K+ position in relation to the interaction around the question he has asked, he also has the epistemic authority to evaluate or assess Stefan's response in turn 6. Stefan also has access to the knowledge needed to answer Tristin's question, but his epistemic status is positioned as less than Tristin's, as it is Tristin who evaluates that knowledge as appropriate. In contrast, Tanya positions her students as K+ and herself as K–, and makes no evaluations of what her students say. These differences relate strongly to the type of question the teacher asks, as well as the purpose of the question being asked. These differences are made visible through how the teacher repeats what the student has said, as this is where the teacher treats the student's contribution as belonging to their epistemic domain, or as something to which they have no access (Kapellidi 2015).

This access to knowledge, and in particular what knowledge can be taken as shared by teachers and students, what knowledge only the teacher has access to at the beginning of the interaction, and what knowledge only students have access to, is a key feature of classroom interactions that both students and teachers orient to. The treatment of epistemic issues, such as access to knowledge, also contributes to students' (and teachers') beliefs about what mathematics is and what it means to do mathematics (Ju & Kwon 2007), which is explored further in Chapter 6.

Epistemic access is not the only epistemic issue that is negotiated or managed in classroom interactions. Teachers and students also treat answers to questions or statements as appropriate (aligning) and affiliative or not. Institutionally the teacher has the right to tell, know, and assert knowledge about the subject matter. In some interactions it can then be treated as

```
44  Tim:        so then, (1.1) we know this. you know that
                those (.) four add up to two hundred and
                seventy-one, (0.4) so then I suppose what you
                could do (.) is say that two hundred and
                seventy one plus the maths mark, well has got
                to equal three hundred and fifty doesn't it.
                (1.2) does that make sense?   [yeah Shane.]
45  Students:                                [yeah       ]
46  Shane:      it could also be seventy one just because the
                question doesn't specify which aver[ age    ]
47  Tim:                                           [hold on].
                we're going back to that, we will go back to
                that. let's do this one. so, what is that
                number.
48  Simon:      seventy nine percent
49  Tim:        seventy nine is it? so by by that sort of
                logic, that there would have to be seventy
                nine percent. seventy nine, at least, because
                seventy nine is the number when that you add
                up the five numbers and divide by five,
                that's the one that gives you seventy
                percent. yeah? is everyone, does, is everyone
                happy with that. so you know when you do the
                mean you do it by dividing, quite often when
                you're solving these (.) mean type problems
                you end up timesing. wha- what we really did
                was say, mystery number divided by five is
                seventy, so what is five times seventy. five
                times seventy is three hundred and fifty so
                that's what the total must add up to. yeah? I
                think we'll wait a few lessons. I think we'll
                try another one like this and make sure next
                time everyone can get it. (1.1) right. (0.9)
                oh what were you going to say
50  Shane:      it could also be seventy one because it
                doesn't specify which average it is.
51  Tim:        right I've got a feeling that Michelle's
                parents au- are not perhaps quite au fait
                with GCSE mathematics. they've just said her
                average test score, and its they haven't
                actually specified whether they're talking
                about the mean, or the mode or the median or
                perhaps some other type of average. okay so
                we didn't actually say the mean average. it
                was just an average. so- oh what are you
                saying it could be
52  Shane:      for the mode it could be seventy one
```

Extract 5-3 Epistemic rights of the teacher.

inappropriate if a student tries to assert the knowledge being discussed. In Extract 5-3, the class have been working on a problem involving averages, and so far the interaction has focused on calculating the mean. In turn 46, Shane offers an alternative answer to the question currently being discussed. However, Tim prevents him from asserting this knowledge (which he does by interrupting him and taking the turn) as he has more to say about the previous answer, and it is Tim's right to do this, as shown by Shane stopping his turn and allowing Tim to continue. Furthermore, whilst Shane gives a brief explanation for his new answer in turn 50, it is Tim who gives a detailed explanation in turn 51.

Whilst Shane's answer is correct, it is not treated as appropriate by Tim. As well as being appropriate for the interaction, students' turns can also be treated as aligning or not, and as affiliative or not. Alignment refers to the structure of a turn; for example, if the teacher asks a question and the student answers the question, then the student's response aligns with the teacher's initiation. Disaligning turns can include interruptions or requests for clarification which disrupt the flow or trajectory of the interaction. Affiliation is a social construct, where agreement is treated as affiliative and disagreement is treated as disaffiliative (Stivers 2008). In classroom interactions, a student response could be both aligning and disaffiliative—for example, when a student gives a different answer to a question that another student has already answered, but this new answer is accepted by the teacher and enables the interaction to continue smoothly.

In this chapter I focus on how knowledge, knowing, remembering, and understanding are treated by teachers and students as they interact in the mathematics classroom. In Chapter 6 I return to this negotiation of knowing and sense making, but with a focus on the consequences on the identities of teachers, students, and mathematics. The analysis discussed here was instigated when, in the transcripts I collected, I noticed that the word 'understand' appeared more rarely than I expected, particularly given the frequency with which teachers talk about teaching for understanding. A crude count of the number of times the words 'think', 'know', and 'understand' appear (including stemmed words such as 'thinking' and 'knows') revealed that think occurred 538 times, know occurred 392 times, remember occurred 111 times, but understand appeared only 50 times. However, this raw count tells us very little about when the words are being used and what is being done with their use by either teachers or students. There are also other ways of talking about understanding without using the word 'understanding', such as 'do you get it?' The distinction between what knowing and understanding mean in the literature, as well as when talking to mathematics teachers, is not clearly delineated, but it is the difference in how these terms are used in classroom interaction and what is being done with these words that is of interest here, not their meaning.

In this chapter we will look at two unusual lessons where the teachers referred to understanding as they interacted with their students. These two interactions are unique for the way in which they talk about understanding, even within the practices of the teachers involved in the other lessons that they video recorded. As such, they are deviant cases. Yet they illustrate two clear practices that say something about the nature of mathematics that the students are working on in these lessons.

Teachers' Use of Understanding

One of the often-stated goals of teaching is to develop students' understanding of the subject. I begin this section by outlining the different ways in which teachers make this goal explicit in mathematics lessons. The first few examples are taken from teacher monologues; that is, where the teacher is speaking and the class are listening, usually from the start of the lesson. For example, teachers can introduce the objectives of their lessons as being about understanding. Both Tyler in Extract 5-4 and Trish in Extract 5-5 introduce the purpose of the lesson explicitly in terms of understanding.

But understanding can also be the purpose of different tasks or activities. In the next two extracts, Extract 5-6 and Extract 5-7, Tanya and Theresa describe one of the goals of a task as understanding.

The first three of these extracts all occurred in the first few minutes of the lesson, where either the lesson is being introduced or the first task of the lesson is being introduced; this is the most common place to find understanding explicitly mentioned in the context of being the aim or goal of a lesson or task. Only Extract 5-7 offers an example of a later task being introduced in terms of a goal of understanding in any of the lessons in the corpus, even though in most lessons there are several tasks that students have to work on.

```
16  Tyler:    …okay so, you're objective today, understand
                that some (.) sequences have limits,
                understand what a limit is, okay and then
                we'll be finding some of the limits of
                sequences and seeing why:, I'll do a problem
                for you first to see why they're important
                (0.6) okay …
```

Extract 5-4 Tyler introduces the objective as being about understanding.

```
19  Trish:    …now you should have done some probability
                last year but today we're just gonna recap
                (.) on probability. make sure we understand
                the probability scale and particularly the
                vocabulary that we associate with
                probability. okay…
```

Extract 5-5 Trish introduces the focus as being about understanding.

There are just two examples where the teacher talks about the purpose of understanding within an activity or task that the students have already been working on for some time. Extract 5-8 occurred in the middle of a period of time where students were working independently on a revision worksheet, and again was not said in interaction with any students.

Similarly, Tracy emphasizes the importance of the role of written mathematics in demonstrating understanding as her students are recording the outcomes of an investigation. Whilst it is a common saying that you understand something if you can explain it, the only references to this are in Extract 5-9 and Extract 5-17

```
4    Tanya:    ...so first off you can just come up with any
                method I just want you to get the question
                and understand it and think about it. and get
                a method down on your boards then I'm going
                to give you- as you're doing that I'm going
                to give you an A3 sheet of paper on each of
                the tables and you've got to agree a method
                you think that no other group will have come
                up with...
```

Extract 5-6 Tanya explains the aim of the task is to understand it.

```
141 Theresa:   ...if you have one in front of you, please look
                at it but we'll actually just do one on the
                board and then you can all help each other to
                understand what's going on...
```

Extract 5-7 Theresa introduces new task with a focus on understanding.

```
149 Todd:      ...if you want to just double check that you
                understand where that came from next time
                you'll see that it will be on the exam papers
                so it's worth making sure you can do it like
                that. if you don't understand talk to me or
                somebody else ah about it please...
```

Extract 5-8 Todd asks students to check their understanding.

```
37   Tracy:    ...but, it's the setting out, so you can
                understand and follow what you've done.
                because with like so much maths, we know-I
                know that you understand it, if you can
                explain to me or somebody else what you've
                written and what you've done. okay...
```

Extract 5-9 Thelma states that you understand if you can explain it.

So, understanding can be an aim in mathematics teaching, and many teachers would probably claim that it is their aim or goal if you asked them, but this is not always or frequently made explicit to students during a class. In Extract 5-8, understanding is not just something that is to be gained during the lesson; it is also something that can be assessed. Only eight of the seventeen teachers used the word 'understand', or synonyms like 'make sense' or 'get it' at all, even when sharing the aims of lessons or tasks, and these eight teachers did not do it in all of their lessons.

So far, we have only examined the use of understanding as a goal or aim of mathematics teaching. There are other ways in which teachers talk about understanding during their teaching. The two most common, which I outline next, are to perform understanding checks (Liebscher & Dailey-O'Cain 2003) and in relation to the meaning of specialized mathematical vocabulary.

Understanding Checks

Frequently in many classrooms, teachers ask questions such as 'do you understand?' or 'does that make sense?' These are often referred to as understanding checks. Koole (2010) showed in his research that these understanding checks are usually followed by *claims* of understanding by student(s), not *demonstrations* of understanding. This is also the case in the data considered in this book. Extract 5-10 and Extract 5-11 are examples of these understanding checks, with Extract 5-10 using the word 'understand' explicitly and Extract 5-11 asking if what has been discussed makes sense.

```
39  Tyler:     [okay] that is, does everyone understand that
                idea
40  Students:  yeah
41  Tyler:     so then if this, …
```

Extract 5-10 Understanding check in Tyler's lesson.

```
171 Toby:      … and you take away twenty-four, you can sort
                of count down to sixteen, so it's about going
                down and remembering it's minus sixteen
                ((inaudible)) number. so, does that make
                sense?  is that a good reminder?
172 Student:   yeah.
173 Toby:      yeah?  so, I'll give you a few minutes on a
                few questions, then I'll go back to that
                starting activity and see if you've made
                progress, and …
```

Extract 5-11 Understanding check in Toby's lesson.

Generally, when a teacher does an understanding check it is followed by a claim of understanding, often by students in chorus, as in Extract 5-10. However, sometimes the understanding check does not include the opportunity for a response at all; the interaction continues as if the understanding check had not occurred. One example of this is given in Extract 5-12

This tendency for understanding checks to not be responded to can lead to ambiguity as to whether a response is needed. In Extract 5-13 Trish asks if everybody understands the difference between 1D, 2D, and 3D. This is followed by a brief pause where no student responds, after which Trish downgrades the statement to 'kind of'. Sophia then gives the usual claim of understanding, which Trish acknowledges with a 'thank you', indicating that this is what she saw as being the relevant response to her initial understanding check, and is the response that enables her to continue as she does in turn.

Both these scenarios show that these claims of understanding are used to support the smooth progression of interaction (Krummheuer & Brandt (2001), as cited in Krummheuer 2015; Stivers & Robinson 2006); they enable the interaction, and the lesson, to continue in the way the teacher has planned or intended. These so-called understanding checks are not so much about checking students' understanding but are more about seeking students' assent to continue. They act to close down the current focus of the interaction and move to a new topic.

```
178 Tiffany:  … but if I drew another rectangle and had
              that as two (.) and four, that isn't similar
              to that cus they're not in the same
              proportions, yeah? does that kind of make
              sense? it's a bit of a weird mathematical
              word. right? but we have to get used to it in
              a mathematical thing, so it doesn't quite but
              congruent certainly is going to come up.
              okay. so we're going to start with
              reflections, but…
```

Extract 5-12 An understanding check with no opportunity for a response.

```
38   Trish:    so we've got length and width and height.
               okay. so that's good. everybody understand
               now what we're talking about when we're
               talking about dimensions. everybody happy
               with that. everybody understand the
               difference between one d, two d and three d?
               (.) kind of.
39   Sophia:   yeah I do
40   Trish:    thank you Sophia
41   Stephen:  so do I
42   Trish:    okay what I'd like you to do now…
```

Extract 5-13 Does an understanding check require a response?

The Meaning of Specialized Vocabulary

The most common use of the words 'understanding' and 'meaning' in whole-class interactions is in the teacher's questioning about the meaning of particular mathematical words. In Extract 5-14 a student (Shannon) has (correctly) used the word 'isosceles' in their explanation for why two angles inside a triangle are equal. Todd picks up on this word use and asks them what they understand by the word 'isosceles'.

Shannon's explanation of the meaning of isosceles then includes the word 'perpendicular', this time used inappropriately, which Todd subsequently follows with a request for the meaning of perpendicular. This practice of focusing on understanding or knowing the meaning of specialized vocabulary appears to be teacher dependent. Four of the teachers use the words 'sense', 'understand', or 'meaning' in relation to vocabulary in all their lessons, whereas the other thirteen teachers do not use these words in relation to vocabulary at all. In Extract 5-14 this understanding or meaning is associated with their use in explanations; the learning or knowledge of the word is not the pedagogical aim of the interaction, though it may be a goal of the embedded exchange. Todd also talks about the meaning of procedures. In Extract 5-15 Skye has explained why a particular answer is wrong by reciting the rule that 'a minus and a minus makes a plus'. Todd initially accepts this rule as an explanation, until turn 93, where he asks for the meaning of the phrase. This particular rule is contentious within mathematics teaching, as it works as a useful mnemonic for many cases, but there are also situations where the rule does not work.

Todd is the only teacher who talks about understanding, meaning, or sense in relation to procedures or processes, and he only does it in situations like the one in Extract 5-15, where the procedure or rule described by a student is one that is potentially contentious.

```
12  Todd:      [ah] that's a good word isn't it isosceles.
                1hat do you understand by isosceles
13             (0.9)
14  Shannon:   erm two lines that are (0.4) perpendicular
                (they're straight lines) and then they have
                two of the same angles
15  Todd:      we're getting lots of different mathematical
                words aren't we. um two lines are
                perpendicular what do you mean (.)
                perpendicular.
```

Extract 5-14 Understanding the meaning of isosceles.

88	Todd:	I won't ask you to put your hands up but a lot of people did that didn't they and why's that wrong? (1.0) why's that wrong.
89		(4.7)
90	Skye:	because a minus and a minus makes a plus
91	Todd:	minus and a minus make a plus. so what do I need to change.
92	Skye:	it's plus three
93	Todd:	right. so that's a plus there. um (.) there's various ways of understanding that. when she said a minus and a minus what was she () doing it to. in what sense is it a minus and a minus ((clears throat)) Sam.
94	Sam:	er ((inaudible))
95	Sean:	minus [times]
96	Todd:	[so this] negative three and this negative q are being multiplied making positive three q. that's a good example of a mistake because it's something very easy to do and of course that's why I put the two minuses in the question to try and trick you into doing that. which worked (.) to some extent.

Extract 5-15 Understanding the meaning of a procedure.

There is variation in how the different teachers treat mathematical language and mathematical procedures. Whilst only a few teachers talk about the meaning of language, others do discuss vocabulary, but largely by defining it or using it rather than discussing it with their students. In these contexts, the definitions are talked about in terms of knowing them and remembering them, rather than understanding them. Similarly, procedures are talked about in terms of knowing and remembering, as well as using and applying, in all except Todd and Tess's classes. This reinforces the view of mathematics as a set of facts, definitions, and procedures that need to be remembered, rather than as a sense making activity.

Next, we will look at two unusual lessons where the teachers referred to understanding as they interacted with their students. These two interactions are unique for the way in which they talk about understanding, even within the practices of the teachers involved in the other lessons that they video recorded. As such, they are deviant cases. Yet they illustrate two clear practices that say something about the nature of mathematics that the students are working on in these lessons.

Two Unusual Teacher Uses of 'Understand'

In this section I turn to considering two unusual, in the sense that they are rare, ways in which teachers use the language of understanding in their teaching. The first of these is from Trish's lesson on arithmetic with negative numbers. The students have been given a well-known published task (https://www.stem.org.uk/rxu56) where they have been given some cards which include some temperatures in some famous cities around the world, and some journeys between these cities, and they need to work out the change in temperature as they move between two cities. Some of the city temperatures are given and some are missing. Likewise, some of the temperature changes are given and some are missing. From combining these cards, it is possible to work out the temperature in every city used and the temperature change for every possible journey.

In the discussion above about using understanding to describe the aims or objectives of a lesson or a task, we have only considered teacher monologues. Trish also uses understanding as an aim for explanations when she and her students are interacting. The students have been given some time to work in pairs to find out the missing temperatures and temperature changes before Trish then leads a whole-class discussion to go through the solutions. Throughout this interaction Trish makes it explicit that the students need to explain their answers. Sometimes a report of the calculation they used to get their answer is sufficient, but on occasions Trish asks for more detail, as can be seen in Extract 5-16. In Extract 5-16 Trish is asking a student to explain in order to help her and other students to understand.

Earlier in the same lesson, Trish explicitly outlined the purpose of explaining after a student had complained that 'it's hard to explain', shown in Extract 5-17.

In this one lesson students are asked to explain how to use negative numbers, with a justification that it helps both them and their peers to understand. Here Trish treats the role of explanations as one of supporting or developing understanding.

```
234 Sarai:    is it er plus five degrees
235 Trish:    right explai:n it to me so that I understand
              but also (0.3) so that people that haven't
              done this one understand (.) cause I know
              that th- there's quite a few people who
              haven't done it…
236 Sarai:    is it because (.) Rio is forty degrees and
              (.) I've worked out that Khartoum is thirty
              five degrees you just add five
237 Trish:    right (.0.7) where did you get thirty five
              from for Khartoum, was that one of the ones
              that we've worked out:
238 Sarai:    yeah
```

Extract 5-16 Explain so that we understand.

```
231 Trish:    but it is hard to explain. that's the point
              of doing this because it's to get you to
              understand (0.4) how to use negative numbers.
              okay so it's really really important, it's
              one of the big things that when you do
              algebra or when you do (.) ah many many
              others topics with maths and in other
              subjects as well, when you're working with
              numbers, negative numbers come up a lot. and
              if you can't (.)manipulate them properly
              you're going to get yourselves (.) all into a
              bi- bit of a stew.
```

Extract 5-17 The point of doing this is to get you to understand.

```
291 Sam:      well you've got to work them out (1.2)
              factoring in the distance from the (pivot)
292 Tess:     I don't understand what them is.
293 Sam:      all the um points of forces (1.1) moments I
              suppose
294 Tess:     all the moments
295 Sam:      yes
```

Extract 5-18 I don't understand your explanation.

In this final example of how teachers can use understanding in their interactions, Tess uses a claim of not understanding to get her students to develop the specificity of their explanations. The lesson is on mechanics and is focusing on moments.

This use of a claim of not understanding in Extract 5-18 also reinforces the idea of an explanation leading to or demonstrating understanding. In this case, Tess is stating that the explanation is not sufficient for her to understand, implying that the purpose of the explanation is for her to understand. It is perhaps surprising, given that the purpose of explaining in mathematics lessons, by teachers and by students, is to support understanding, that this purpose is so rarely made explicit in classrooms.

Students Claiming and Demonstrating Understanding

For what follows, the distinction between *claiming* and *demonstrating* understanding is useful. It was Harvey Sacks (1992) who first made this distinction between *claiming* and *demonstrating* knowledge or understanding. A statement such as 'I understand' or 'I don't know' in itself does not tell us anything about what is or is not being understood or known. These are therefore *claims* of understanding or not knowing. Demonstrations, on the other hand, tell us something about what is known or understood. Koole (2010) describes this distinction as the difference between showing *that* you have understood something and showing *what* you have understood. This distinction is crucial in identifying or assessing students' knowledge or understanding of the mathematics being learnt.

Claims and demonstrations of understanding are treated differently in classroom interactions. A claim of understanding supports the smooth progression of interaction, is affiliative, and reinforces the idea of shared knowledge and understanding of the topic of discussion (Weatherall & Keevallik 2016). The most common occurrence of a claim of understanding follows a teacher's understanding check; as, for example, in Extract 5-10 and Extract 5-11. However, by the very nature of the claim there is no information about what the students understand. Claims of understanding can also occur outside of an IRE sequence, as in Extract 5-19.

Whether the teacher has invited the claim of understanding or not, the interaction continues without disruption, deviation, or interruption; that is, these claims are treated as if the aim of students understanding what has gone before has been achieved, without necessarily any sharing of what has been understood or how many students within the class have understood. This can potentially be problematic, as understanding checks are yes-preferred actions, so students will avoid the dispreferred response: a claim of not understanding. This resonates with the treatment of mistakes discussed in Chapter 4. Understanding is needed in order for the teacher to progress the lesson, yet there are few opportunities for students to indicate that they do not understand and making claims of not understanding is dispreferred.

The situation with student demonstrations of understanding is different. Student demonstrations of understanding often follow the teacher explicitly asking students a question which requires them to demonstrate their understanding. In these situations, the interaction that follows focuses on the content of these demonstrations of understanding and is consequently extended. Examples mentioned above include Extract 5-16 and Extract 5-14, but also Extract 5-20 below.

Demonstrations of understanding by students are followed by the teacher asking a follow-up question which focuses on the content of what the

```
145 Sophie:   I get it now
146 Trish:    good (1.3) okay (0.6) what about if we were
              working out (1.3) erm we've done Wellington,
              we've done Anchorage, Khartoum. who's done
              Khartoum. (1.7) Simon.
```

Extract 5-19 Student unsolicited claim of understanding.

```
68  Tim:      yea:h it is because of that so why, why does
              that make sense, can you say so why does that
              makes sense to use four (in the numbers.) it
              is something to do with the fact that it's
              quarterly. Sayed?
69  Sayed:    because there's four quarters
70            (1.1)
71  Tim:      where
72  Sayed:    in the end, (in a year)=
73  Tim:      =yeah there's four quarters in a year but,
              so, yunno so …
```

Extract 5-20 Student demonstration of understanding.

students have demonstrated. These explicit requests, combined with the follow-up within the IRE, still contribute to the smooth progressivity of interaction, but the progressivity of the interaction is determined by the teacher's initiation. So, both claims and demonstrations of understanding contribute to the smooth progression of classroom interaction. This is not the case when there are claims and demonstrations of not understanding.

Students Claiming and Demonstrating Not Understanding

Student claims and demonstrations of not understanding in whole-class interactions are extremely rare, but also do not always occur within an IRE sequence. Here I am making the distinction between knowing and understanding where demonstrations of not understanding use the language associated with understanding in either the teacher's initiation or in the student's turn. Demonstrations of not knowing in terms of giving incorrect responses have been discussed in more detail in Chapter 4. In my data there are only four demonstrations of not understanding and two claims of not understanding in total. When there is a claim or demonstration of not understanding, in all but one of the cases the interaction is briefly interrupted, but the teacher 'parks' the issue until later in the lesson rather than disrupting the topic progression.

The rarity of demonstrations of not understanding is possibly a consequence of the difficulties of articulating what it is that you do not understand, but also the public nature of whole-class interactions, given that the student usually initiates the topic of the interaction. Claims and demonstrations are often dispreferred actions, and in many cases are disaffiliative in nature, challenging the main purpose or aim of the interaction itself.

```
173 Tristin:   numbers pick one of the numbers as a value
               for n. work out the missing numbers, using
               the same shape, apply that rule to somewhere
               else, see if you can add them up to be um
               using this rule like this one if I draw that
               C anywhere else I should find that I get er
               five n add two will tell me, if I draw a C in
               here, er draw this c in one two three four
               five. five times thirty-two is a hundred and
               sixty, add two makes a hundred and sixty-two.
               Sean
174 Sean:      I don't understand
175 Tristin:   okay I'll come and see you. it will be seven
               minutes, to have a go, starting now.
```

Extract 5-21 A student claim of not understanding.

In all cases the teacher treats the issue of understanding as an individual issue specific to the student who has made the claim or demonstration. The claim is made publicly during the whole-class interaction but is dealt with privately on a one-to-one basis later on. The responsibility for understanding is placed with the student. The teacher's instructions or explanations are treated as sufficient for the remainder of the class, and the issue is with the one student who made the claim. The follow-up on an individual basis also emphasizes the goal of understanding, as the teacher and students both treat this as something that needs addressing and dealing with. However, a claim or demonstration of not understanding also treats this goal as not being achieved, which is a disaffiliative act. The teacher minimizes the disaffiliative nature of a demonstration of not understanding by making it about this one student.

There is one exception in the data, where the teacher, Todd, deals with the demonstration of not understanding immediately and publicly, but in this situation he treats it as an issue of not remembering rather than not understanding. In Extract 5-22 Susie has explicitly stated that she does not understand how to find the equation (of a straight line). Todd begins by pointing to the time where they worked on finding the equations of straight lines as a class, before summarizing what the three and the one in the equation in question relate to.

Turn 76 continues by describing and setting the next question for the class to work on and continues for some time after the part presented in Extract 5-22. This is another way in which this extract differs from the other demonstrations and claims of not understanding. In all other cases the students in the rest of the class have already been set a task to work on by the teacher, and the claim or demonstration occurs after the teacher's explanation of this task but before they have asked the students to begin working

```
74  Todd:    yeh did that come from Sam or from Sophie
             couldn't work it out. yes three x plus one.
             um okay that's fine. who wants to ask
             anything about the first side, you do. go on
75  Susie:   um how do you find out the um equation I
             don't understand it
76  Todd:    no okay. this three ((clears throat)) I
             remember we went to the computer room and
             drew some things didn't we this three turned
             out to be the gradient and this one turned
             out to be where it goes through the y-axis.
             so once you've found out the gradient is
             three you can just stick it in there um that
             requires quite a bit of thought doesn't it um
             (.) at this stage with the exams coming up
             just stick it in there in front of the x and
             you put plus one on the end because of that
             and it all fits in with the pattern we saw
             when we drew lots and lots and lots of these
             in the computer room. um (.) right on the
             back let's speed up a bit. you're getting a
             bit restless, …
```

Extract 5-22 Response to a demonstration of not understanding.

individually or in groups on the task. In Extract 5-22 this task has not yet been given. In the other situations, 'parking' the issue only involves a short period of time before the teacher goes to work one-to-one with the students. Here, however, if the issue was 'parked', there would be a longer period of time before the issue could be dealt with on an individual basis, as the teacher would need to give the other students something to do first. Todd has invited the students to ask any questions, and this demonstration of not understanding functions as a question about where the equation came from. Todd's explanation answers this question but his hedging within the explanation 'that requires quite a bit of thought doesn't it um (.) at this stage with the exams coming up just stick it in there in front of the x and you put plus one on the end' suggests that he may not be dealing with the issue of not understanding. He has given a rule that works, but references the time they spent working on why the rule works rather than explaining it immediately.

Students Claiming 'I Don't Know' or 'I Don't Remember'

If students cannot answer a question, they have several ways of demonstrating this. The most common response is 'I don't know', but there are other options such as 'I don't understand' or 'I don't remember' (among others, but here I am focusing on those responses that refer to cognitive processes). We cannot know from these utterances whether a student does or does not know or understand, but these utterances do more than just declare a cognitive state. They also serve an interactional purpose. Each one shows that the student understood that it was their responsibility to answer the question, as they are giving a response to the question, just not necessarily the response the teacher wanted or was expecting. Yet whilst this response of 'I don't know' is a response to the teacher's question, teachers generally do not treat it as an acceptable response in their turn that follows. That is, the teacher treats the question as something the students should know the answer to. Teachers instead persevere with the question in a variety of ways.

Immediately before the start of Extract 5-23, Simon has given 5°C as the difference in average temperature between two cities. Trish then asks whether the difference is plus or minus 5°C when you are travelling in a particular direction. In turn 205, Simon hesitates and pauses, before saying 'don't know', and then emphasizing this by repeating it with the addition of the word 'really'. Trish's question is a binary question: the choice is between two specific given options, +5°C or −5°C. Simon chooses to say that he does not know the answer rather than choosing one of these options. Trish then repeats her question, beginning by reinforcing that Simon's original answer of '5°C' was correct. Simon offers a hesitant answer—indicated by the delay in answering, the hesitation marker, and phrasing his answer as a question—in turn 210, which chooses one of the options. This response is followed by another pause before Trish invites Simon to answer again by directly referencing the fact that the question is binary, and consequently even if he guessed, he would have a fifty per cent chance of being correct. Trish does not acknowledge either Sarah's or Simon's responses, and after another pause Simon changes his answer to −5°C, again phrasing his answer as a question. Trish's next turn is immediate and asks a new question.

```
204 Trish:     so the gap is five degrees (.) is it (.) plus
               five degrees (.) or minus five degrees
205 Simon:     er (1.7) don't know (.) I really don't know
206            (3.7) ((Trish is writing on the board))
207 Trish:     it is five, you're right (0.7) but is it plus
               five or minus five
208 Sarah:     °minus°
209            (1.5)
210 Simon:     er is it plus five
211            (1.3)
212 Trish:     you've got a fifty fifty shot Simon
213            (1.4)
214 Simon:     is it minus
215 Trish:     why is it minus
```

Extract 5-23 Trish pursuing an answer following 'I don't know'.

Whilst Simon has demonstrated in his initial response in turn 205 that he needs to answer the question, Trish does not treat the response as appropriate, and in turn 212 encourages Simon to guess. This pursuit of an answer despite a student stating that they do not know the answer does not only arise in situations where it is possible to guess an answer. In Extract 5-24 Tyler is asking for an explanation for why the question is about indirect proportion, and in turn 281 Scott shows that he is aware that the response to Tyler's question in turn 280 should be an explanation, as he begins his turn with 'because', but then he pauses before saying he doesn't know. Similarly to Trish in Extract 5-23, Tyler responds to this by first confirming that the answer Scott gave of 'indirect' is correct, before re-asking the question. Following another pause of 1.7 seconds, Scott then gives the required explanation.

In Extract 5-23 and Extract 5-24 the teacher has pursued an answer even when the student has stated that they do not know the answer. The teacher turns between the student saying they do not know and the student giving the answer the teacher is looking for do not contain any additional information or knowledge that might enable the student to answer the question. This suggests that when teachers ask known-answer questions, they are positioning both themselves and their students as having access to the knowledge which answers the question. By saying 'I don't know', the student is challenging this K+ position by positioning themselves as not having the access (K–). The teacher is not treating the student's statement of 'I don't know' as a reference to the cognitive state of not knowing, or, to put it another way, as the student being in a K– position. In Extract 5-25 the teacher Tom does add an extra piece of information.

In turn 50 Sophie states that she does not know the answer. Tom then repeats his explanation from turn 49, inserting 'to keep the fraction the same' and changing the wording of what needs to happen to the denominator. Sophie then immediately and appropriately answers the question in turn 52,

```
278 Tyler:    ...at eight miles an hour a boy takes five
              hours how long will he take at twelve miles
              an hour is that indirect or direct
279 Scott:    indirect
280 Tyler:    why
281 Scott:    because (1.2) don't know
282 Skye:     ooh
283 Tyler:    you're right (.) I just want to know why
284           (1.7)
285 Scott:    cus the amount (.) the speed it goes the time
              goes down yeah
```

Extract 5-24 Tyler pursues an explanation.

```
49  Tom:      if she's got three times as many pieces on
              the top, she needs three times as many pieces
              on the bottom to make it the same. so what
              are you going to multiply it by on the
              bottom?
50  Sophie:   u::m, I don't know.
51  Tom:      okay, let me explain again. if she's got
              three times as many pieces on the top, to
              keep the fraction the same she needs the
              bottom to be three times as big as well.
52  Sophie:   three.
53  Tom:      So you times it by three?  what happens?
```

Extract 5-25 Tom pursues an answer by re-explaining.

```
181 Trish:    right there is a ten gap between forty and
              thirty, but if I add forty and thirty I get
              seventy so why (.) why don't I just stick the
              minus at the front. (2.4) Seth?
182           (2.4)
183 Seth:     I- (0.3) I can't remember
184 Trish:    no Sarah?
185 Sarah:    is it minus ten because (1.1) when, when it's
              in minus numbers you take away get (.) like a
              higher number (.) so if it's minus ten then
              you take away minus thirty then it gets to
              minus forty which is the temperature range,
              (Verhoinsk)
```

Extract 5-26 Trish responds to a student claim of not remembering.

which is accepted by Tom in turn 53. Tom's rephrasing makes clearer what needs to be the same, but treats Sophie as knowing what needs to be done to the denominator in order for the fraction to be equivalent, despite Sophie's claim of not knowing.

There is a difference in how the teachers treat student claims when the claim refers to not remembering or forgetting something. In these cases, the teacher redirects the question to another student. In Extract 5-26 Trish initially asks Seth, who, after a long pause of 2.4 seconds, hesitantly states that he 'can't remember'. Trish then passes the question to Sarah, who gives the explanation that Trish was looking for.

In Extract 5-27 the student states first that he does not know, before stating that he cannot remember. Again, in this situation the teacher redirects the question immediately to another student.

In Extract 5-28 Tara treats Susie's turn as an issue of remembering, even though Susie has stated that the issue is one of knowing. Again, the teacher does not return the question to the student, as was the case with a claim of not knowing, but rather partially repeats the question before asking another student.

In these scenarios the student is not directly challenging the teacher's positioning of them as having access to the needed knowledge. By saying 'I can't

29 Thea: a diagonal. now, we were talking, I think it
                was you and Sarah, about it was a diagonal, a
                wonky line, you said a wonky line. now, what
                do you feel like was wrong with that?
30 Shane: I don't know (.) I can't remember.
31 Thea: Salma?
32 Salma: straight like that or like that.
33 Thea: So, does being a diagonal, a ruler, they're
                all dead straight. So when you see those
                lines, I think a lot of you think like that,
                don't you?

Extract 5-27 Thea redirects the question following 'I can't remember'.

58 Tara: absolutely, yeah, it mentions volume, doesn't
                it, so that suggests that the volume is the
                topic that it _is_, and yes, clearly, it's
                right-angled triangles but yes, it was a
                volume question, it wasn't a Pythagoras
                question. and what were the, what were the
                mistakes on this? Susie, what was one of the
                mistakes on this?
59 Susie: um, (0.9) I don't know.
60 Tara: okay, who can remember? I'm not putting that
                paper in front of you because we're still
                doing the early one but two things wrong with
                this. one of them is (1.3) Sheila?
61 Sheila: I don't know.
62 Tara: Sasha?
63 Sasha: is one of them to do with cubes?

Extract 5-28 Reframing 'I don't know' as 'I don't remember'.

remember', they are saying they had or have access to the knowledge but have forgotten it at this point in time. This claim of not remembering has not challenged the assumption the teacher has made of this knowledge being shared, and the teacher continues to treat the knowledge being shared by redirecting the question to another student. On the other hand, this assumption that the knowledge is shared is challenged by a claim of not knowing, and is consequently disaffiliative.

The difference between claims of not knowing and claims of not remembering also highlights the differences in epistemic responsibility between students and teachers in each situation. When a student claims that they cannot remember, their response is affiliative in that they are accepting the responsibility for not giving an aligning answer. In contrast, by making a claim of not knowing, students are putting the responsibility for the knowledge on the teacher and are challenging the assumption that the knowledge is shared. The institutional role of the teacher includes responsibility for ensuring that their students have access to the knowledge they are responsible for teaching (Heinemann, Lindström, & Steensig 2011).

Expertise

The expertise or the identity of expert is commonly associated with the identity of teacher, but teachers can orient to this identity in different ways. In
Extract 5-29 the identity of expert is made relevant through the references to
examinations and textbooks, by the reminding of facts and procedures and
the evaluation of students' answers.

Here Tim makes explicit that he knows the types of questions his students
will be asked in their national examinations and in the textbook, demonstrating his knowledge of the curriculum. He also demonstrates his pedagogic
content knowledge by stating that the mean is harder.

Students also orient to the teacher's identity as expert. For example, they
can phrase their answers as questions or their response contains other

```
70  Tim:        … sometimes in the exam they won't give you
                that extra column they'll just give you these
                two, and they'll expect you to know (.) that
                it might be useful (.) to put this extra
                column on, do you know what I mean. and in a
                minute, when you do some practice from the
                text book it's the same thing. they just give
                you this bit of the table and they expect you
                to use your initiative (.) to draw in the
                extra column to do it. okay. well let's go
                through these then, the mo:de, the median,
                the mean and the range. I think we'll leave
                the mean till last because it's a bit like
                the mean one. um Sarai and (.) Sam, paying
                attention now specially, right any offers
                anyone for telling me what, m why of course
                we always want to know why (.) what the mode,
                the median the mean and the range are. (1.7)
                and e- I'm especially interested in people
                answering who haven't answered who haven't
                said anything in class (.) you know for the
                last, last lesson or so cause it's quite
                often it's a bit like the same hands (.)
                going up. those people clearly have no
                understanding. some other people. Sophia?
71  Sophia:     er um days absent three is the mode because
                it's the most common one.
72  Tim:        right. the mode, so these are all days absent
                there's some people won't never had a day
                absent, some people have one day, some people
                had six days, some people had seven days. the
                most common number of days to have like
                absent the mode is three because a hundred
                and twenty-five people had three days off.
                that beats any o- any other sort of number of
                days off, so the mode is three. good choice
                of where to start, well done. ↑u:::m (0.4)
                >go on then< Scott.
```

Extract 5-29 Tim orienting to the role of expert.

markers of uncertainty, as Sophia does in turn 71 in Extract 5-29. By doing this, students are obliging the teacher to make evaluations of their turns. This makes the teacher's expertise in making these judgements relevant, as well as the students' own positions as lacking expertise to make these judgements themselves. Expertise here is about knowledge. Tim has the knowledge and can use it to make evaluations, and Tim's question–answer adjacency pairs are focused on checking whether his students have acquired this knowledge.

Expertise can also be about mathematical behaviours. In Extract 5-30 and Extract 5-31 the teacher models a problem-solving process, in that the task is structured so that initially the pupils are trying out the first few examples (specialising), before making conjectures about what would happen next,

```
18 Tyler:      … okay I want you to look at that (.) first
               question is quite an easy one, the second
               question we have to need to think about in
               terms of (.) what it actually means, (1.3)
               okay and I want you to try your best and try
               and understand (.) how far you can get it
               done, okay. here is your problem. have a go
               at this (.) I've just inherited twelve
               thousand pounds, (0.4) okay and being the
               generous man that I am I want to donate (.)
               some of that to charity (.) but because I'm
               not totally generous, (1.2) okay I'm going to
               donate one quarter of the twelve thousand
               pounds, then the following week I want to
               donate a quarter of that amount, following
               week a quarter of that amount okay how much
               will I donate in each of the first four
               weeks, the first few are obviously easy. how
               much will you donate in total okay let's just
               do the first one together, in week one how
               much have I donated?
19             (1.1)
20 Sasha:      thre[e thousand]
21 Shannon:        [three thou]sand
22 Tyler:      three thousand pounds (3.3) wee:k two: how
               much am I donating if I'm donating a quarter
               of that (.) Sean?
23 Sean:       seven point s-, seven (.) point five, no
               seven hundred and fifty
24             (1.5)
25 Tyler:      okay a quarter of that, seven hundred and
               fifty pounds (.) okay (.) I want you to try
               and work out (0.8) the next (0.5) two weeks,
               and then I want you to think about (0.3) how
               much are you going to end up donating in
               total (0.9) okay. we'll talk about that more
               in a minute. so (0.6) give you two minutes,
               how much are you going to donate in the first
               four weeks, you've got two more to work out.
               talk amongst yourselves, how much am I going
               to donate in to:tal (0.7) off you go
```

Extract 5-30 Tyler models a problem-solving process.

before making a connection between the image presented by a triangle and the original numerical problem, and the mathematical focus of the lesson.

In Tyler's lesson, these mathematical behaviours of specialising, conjecturing, and connecting are not explicitly talked about or mentioned. Thus, whilst the students are contributing to these behaviours and acting in mathematical ways, they may not be aware of this.

Expertise as modelling mathematical behaviours can also be made explicit. In Extract 5-32, Todd explicitly talks about how multiple examples are not sufficient for him to be convinced. This discussion then leads to the students being required to convince Todd that the answer is always a multiple of five, which by the end of the lesson is done by representing the relationships between the numbers in the grid algebraically (Figure 5-2).

```
63  Tyler:    … ↑this was on the corner of the boa:rd (2.0)
              o↑kay, (0.) -↑this was on the corner of the
              board >because this is actually< a useful
              wa:::y (.) of you actually looking at it (.)
              imagine ↑that's my money (1.2) o↑kay, (.) a
              qua::rter of that I'm going to throw awa:y
              and donate, (.) this is my ↑quarter
              (0.6)((shading in middle triangle)) I've just
              given ↑tha:::t (0.4) away.
```

Extract 5-31 Tim connecting problem to an image.

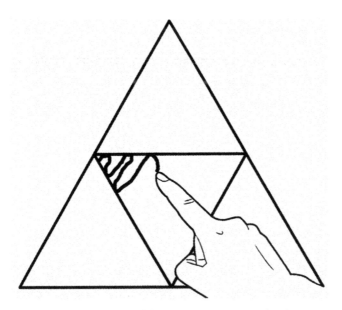

Reprinted from Ingram, J. (2014). Shifting attention, *For the Learning of Mathematics,* 34(3), p. 21 (reprinted by permission of the publisher (FLM Publishing Association: https://flm-journal.org).

Figure 5-1 Triangle representing donation of money.

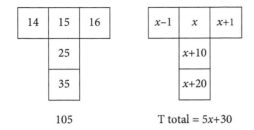

Figure 5-2 T-totals task.

81	Todd:	some people wrote out a few of these and said they all seem to be multiples of five. I can't remember which people, who came up with some sort of observation like this, yes people over there did and other people said they all seem to be multiples of five, and I said will they always be multiples of five, and so they did another one, and they did another one, then they did another one and then they got bored of doing them and said look there's always going to be a multiple of five. um (0.3) but I think that that's not completely convincing, that doesn't completely convince me (.) when people do lots and lots of examples, why do you think maybe I'm not (0.4) totally convinced by that? (3.8) they did another one. I can't remember where it was. they put it somewhere else on the grid it was it was (.) a different number but it was still a multiple of five. and then did another one, another one erm (1.1) wh- why was I not totally convinced by that do you think. (0.8) Seb
82	Seb:	you could turn the (.) t like (0.3) sideways or something and [try then]
83	Todd:	[mm:] you could do, but even if you did and even if you kept it that way round, I still wasn't totally convinced, they would have been convinced and they could have, I don't think they really see, saw what I was make such a fuss about
84	Susie:	((inaudible)) and only one in the whole grid that's like
85	Todd:	yes, only one in the whole grid might be (0.6) different or s[omething]
86	Susie:	[yeah]
87	Todd:	that's what you're saying isn't it. um and that's my problem, that life's full of exceptions isn't, just because something happens a lot doesn't mean it happens always. yeh? um like (0.3) can you think of any examples of that. where you have a rule that doesn't always work.

Extract 5-32 Todd models what it means to be convinced in mathematics.

In these first three extracts, all three teachers have oriented to the identity of expert. All three teachers ask the questions and evaluate their students' responses, but the nature of this expertise is different between the three extracts. In Tim's lesson, being an expert is about using knowledge of mathematics and the knowledge of school mathematics or the curriculum as considered in examinations and textbooks. In Tyler's and Todd's lessons, being an expert was more about behaving and acting in mathematical ways.

One of the most distinctive ways in which expertise is treated in classroom interactions is in the nature of the questions teachers ask. Questions can be known-answer questions (Mehan 1979b), as they are in the examples above, or they can ask for what the students are thinking or for their ideas, as in Extract 5-33. In these situations, it is the students who have the expertise or knowledge, as the questions are about what they are thinking. Whilst there might be a mathematically 'correct' or desired answer to these questions, and the aim of the interaction or the lesson may be for students to develop this meaning or understanding, within the IRE structure surrounding this question the teacher does not usually make an evaluation of whether the response given is correct or not. The 'E' move does not include an evaluation but a follow-up move.

Revoicing is another widely discussed teacher move described in Chapter 3 that influences the relative expertise of the teacher and the students. O'Connor and Michaels (2019) describe revoicing as the echoing or rephrasing of a student's response in a way that allows the student to affirm or contest the teacher's revoicing. Thus the student is treated as the one with the expertise to make the evaluation of what is being said, rather than the teacher.

In turns 33, 35, and 37 in Extract 5-34 Tim is revoicing the response Steven has given in turn 29. Tim attributes the idea that the five numbers need to add up to 350 to Steven, and Steven is given the opportunity to affirm Tim's revoicing in the intervening turns. It is Steven who has epistemic responsibility for what is being discussed.

1	Todd:	… what do you understand by the idea of (.) proof. mathematical proof. p r double o f. (0.9) what do you understand by that (.) concept, that idea. maybe say one thing about it (.) then let somebody else say something else. um:: hands going up. Seth.
2	Seth:	you can't prove anything apart from maths because it's all point of view.
3	Todd:	oh I see (0.7) um: (.) can you give an example or something

Extract 5-33 Todd asks students what they understand by the word proof.

```
29  Steven:   um you need, if yo-, you can find th- like
                (0.3) all the numbers, the end mark, the end
                (.) percentage means that there's like three
                hundred and fifty percent altogether so if
                you divide by five it comes up to seventy.
30  Tim:      right hold on a sec. (0.6) three hundred and
                fifty percent, (0.2) er I suppose, can you
                add percentage together and then get three
                hundred and fifty per
                [cent I suppose so ]okay
31  Steven:   [no what we        ]
32            (1.0)
33  Tim:      so you're saying that if you've got five
                numbers (0.4) and you want to get a mean
                (0.3) of seventy
34  Steven:   yes=
35  Tim:      =those five numbers must add up to (0.7)
                [   w]hat
36  Steven:   [thr-] three hundred and fifty
37  Tim:      three hundred and fifty.
38  Steven:   ye:[h
39  Tim:          [okay] that is, does everyone understand
                that idea.
```

Extract 5-34 Tim revoices Steven's answers.

Expertise is more than what knowledge a teacher or a student has. Expertise is co-constructed, locally contingent, and evolves within interaction. Expertise also has different meanings and roles as we move from one classroom to another. This has important consequences for the mathematics students are learning and beyond.

Conclusion

In this chapter I have examined the interactional contexts in which teachers and students talk about knowing, remembering, and understanding and how these are handled differently in interaction. These differences have consequences on the negotiation of knowledge, epistemic access, and epistemic responsibility.

Understanding is treated by some teachers as relevant to the mathematics classroom. They do this through the goals or objectives they share but also by checking understanding before moving on to a new topic. Yet whilst it may be a goal of lessons, tasks, or interactions, it is not treated as something to be negotiated, challenged, or even to be discussed. There seems to be one exception to this. The meaning of specialized vocabulary is treated by teachers as an issue of understanding. The meaning of individual words is something to be discussed and negotiated.

Knowing and remembering are different. Claims of not remembering and claims of not knowing do different things. The interactional treatment of knowing or remembering is situated within the interactional context of the question or task initiated by the teacher. The way that teachers handle them tells us a great deal about the purpose of the initial question. These claims are made to known-answer questions. These have been portrayed in the literature as testing questions, or questions teachers use to assess students' knowledge. Yet if this were the case, then an answer of 'I don't know' would give the teacher the assessment information they need. Instead, the teacher pursues an answer, even if this means turning the interaction into a 'guessing game'. This suggests that one purpose of these questions is to make the knowledge that the teacher has assumed to be shared public, which then enables this knowledge to be built on in future turns. Claims of not knowing in response to these questions challenge the teacher's assumptions about students' epistemic access, and teachers and students do interactional work to mitigate the disaffiliative nature of this challenge, and to shift the epistemic responsibility. In contrast, claims of not remembering do not challenge this assumption, and are treated as acceptable responses by teachers.

The rarity of claims of not understanding compared to claims of not knowing suggests that whole-class interactions are dominated by discussions around knowing and knowledge, rather than understanding. This is supported further by the few contexts where teachers use the language associated with understanding, compared to knowing, as well as how few teachers use the language associated with understanding in contexts other than understanding checks, which have an interactive and social purpose, rather than actually checking understanding.

This is not the full picture, however, as it is difficult, if not impossible, to distinguish between demonstrations of knowing and demonstrations of understanding in interaction. Even where a teacher's question specifically asks what a student knows or understands, the response does not have to match in type. A common question that teachers ask is 'why', but in mathematics this question is just as often answered by a description of the procedure or process used to get an answer as it is by referring to any underlying conceptual idea or relationship (Ingram, Andrews, & Pitt 2016). Interactionally, there is no difference between how demonstrations of knowledge and demonstrations of understanding are treated by teachers. Perhaps this is what lies behind the difficulties teachers and researchers have in distinguishing between and defining knowledge and understanding.

6

Doing Mathematics

This chapter turns to focus on the subject content of classroom interactions, specifically the mathematics. What does it mean to *do* mathematics in the mathematics classroom, in interaction with the others in the classroom? What are the roles of teachers and students in this doing of mathematics? Or, in other words, what are the roles and identities of mathematics teacher and mathematics student as they are co-constructed in classroom interaction?

It is through mathematics classroom interactions that students learn what it means to do mathematics. Language is constitutive (Ingram 2012; Yerrick & Roth 2005), the words we use to describe ideas, concepts, or activities creates the meaning that these have for us and those we interact with. Similarly, the words we use to describe mathematics, or the activities involved in working mathematically, constitute what mathematics is. In interaction, it is not only the words that constitute what it means to do mathematics, but also what is done with these words. In CA and other microanalytic approaches to studying classroom interaction, it is these word choices, and turn design more generally, that are of interest. How students experience mathematics and come to know what it means to do mathematics is not only influenced by the tasks and activities they participate in, but also how these activities are described and introduced. It is in interactions that mathematics learning happens, and it is in interaction that the meaning of mathematics itself is developed and negotiated.

This chapter brings together aspects of classroom interaction that influence the mathematics that students experience. I show how the way that tasks and activities are talked about and are talked into being affects what it means to do mathematics, and what it means to be a teacher or a student of mathematics. To begin with, I outline two theoretical constructs from existing literature that I have found useful in making sense of the intimate relationship between interactional contexts and the nature of mathematics that students experience. These are sociomathematical norms and identity. Yet these two constructs do not fully explain how mathematics is discursively constructed through classroom interaction. I then turn to some actions and behaviours that are widely considered to be fundamental to mathematics,

Patterns in Mathematics Classroom Interaction: A Conversation Analytic approach. Jenni Ingram,
Oxford University Press (2021). © Jenni Ingram. DOI: 10.1093/oso/9780198869313.003.0006

communicating using the language of mathematics, problem solving, and argumentation. These three foci are each essential to the development of meaning-making in mathematics, as is shown through the extensive research on each of these foci and their explicit inclusion in curricula around the world. This chapter draws on ideas and analyses that focus on the relationship between mathematics and language, sociomathematical norms, and identities to consider what it means to do mathematics, and how the identities of mathematics teacher and mathematics student are co-constructed in interaction, before extending these same ideas to consider the 'identity' of school mathematics.

Norms and Sociomathematical Norms

Norms are patterns of behaviour that are considered normal in the context that they occur. These patterns can be described as practices or activities (e.g. Moschkovich 2015), or hidden regularities (Wood, Cobb, & Yackel 1993), or patterns of interaction. They are collective expectations, though not necessarily explicit expectations, about what different participants in the classroom can do and say. They are 'the way things are done' (Nolan 2012). The idea of norms is widely considered from a range of theoretical perspectives in education research which I briefly discussed in Chapter 2. Classroom norms have been shown to be intertwined with individual beliefs (Yackel & Rasmussen 2002), with beliefs about what it means to learn or do mathematics considered as 'an individual's understandings of the normative expectations shared by a class' (Ju & Kwon 2007, 268). Consequently, the classroom norms around what it means to do mathematics, constructed through classroom interactions and tasks, influence the development of students' mathematical beliefs about what mathematics is and what doing mathematics involves. This chapter focuses not on beliefs about what it means to do mathematics, but on the meanings mathematics has for students and teachers, as demonstrated in their actions within interactions. However, implicit norms have been shown to be an obstacle for the participation of students with low socioeconomic status in classroom interactions (Vogler, Prediger, Quasthoff, & Heller 2018). Norms within a classroom can also vary depending upon the language proficiency of multilingual students within that classroom (Planas & Gorgorió 2004). Whilst a CA approach is agnostic about teachers' and students' beliefs, it does offer a way of making the often implicit norms explicit.

Much of the research building upon the idea of norms has taken a socio-cultural perspective, where norms are social and cultural and play a central role in shaping classroom practice (Xu & Clarke 2013). That is, norms shape and give meaning to the actions of teachers and students in mathematics classrooms (Hofmann & Ruthven 2018). They can also be explicitly stated, or even consciously developed, by teachers, but they may also be implicitly enacted and taken for granted, and enable the smooth interaction between teachers and students in classrooms. Norms can be considered at different levels, ranging from the idea of cultural norms that pervade the teaching and learning of mathematics in particular countries to the level of the particular mathematics classroom or even lesson or activity within a mathematics class-room, with ethnomethodological approaches focusing on the latter. Yackel and Cobb (1996, 178) take an ethnomethodological approach in their exam-ination of norms in mathematics classrooms and define norms as 'regularities in communal or collective classroom activity which are considered to be jointly established by the teacher and students as members of the classroom community'. This definition emphasizes both the classroom level of norms and the joint construction of norms by both teachers and students considered in a CA approach. Importantly, whilst norms are jointly established, they are also continuously jointly negotiated, adapted, and renewed.

Sociomathematical norms are 'the normative aspects of mathematical dis-cussions that are specific to students' mathematical activity' (Yackel & Cobb 1996, 458). What counts as mathematically different within a classroom where asking for strategies that are different to the ones that have already been shared is a norm that Yackel and Cobb give as an example of a socio-mathematical norm. Many of the habits of mind described by Cuoco, Goldenberg, and Mark (1996) would also be true in other curriculum areas. However, the distinction between social norms and sociomathematical norms is not always so clear (Sekiguchi 2006). What is it that makes an explanation mathematical, rather than just an explanation? We could decide that these norms are based on what teachers explicitly state they consider a mathemat-ical explanation to be (e.g. Levenson, Tirosh, & Tsamir 2009). Some researchers also set criteria for what they themselves consider to be mathem-atical, or not. The boundary between what is mathematical or not is far from clear. In an ethnomethodological approach, what matters is what the partici-pants themselves treat as an explanation as they interact. The majority of the time the distinction between an explanation and a mathematical explanation is not made by teachers or students (Ingram, Andrews, & Pitt 2019).

An ethnomethodological approach, such as the CA one, views norms as continually negotiated and co-constructed in interaction. Furthermore, participants, norms, and contexts are mutually shaped through interaction. Whilst classroom and sociomathematical norms are actively negotiated by teachers and students, they may not be conscious of these norms until they are breached (Herbst & Kosko 2013; Heritage 1984). Norms are an 'accountable moral choice' (Heritage 1984, 76), rather than behavioural constraints or rules. The majority of classroom interactions run smoothly because of the taken-for-granted norms that allow us to make sense of each other as we interact, despite the fact that classroom interactions are complex and involve many different individuals. This perspective on norms also shifts what is meant by 'participation' in mathematical activity. From a sociocultural perspective, participation often refers to the idea of taking part in a 'community of practice' (Lave & Wenger 1991) over time. Conversation analysis draws on the use of participation by Goodwin and Goodwin (2004), where participation is a micro phenomenon. It is achieved in the moment-to-moment flow of interactions and 'refers to actions demonstrating forms of involvement performed by parties within evolving structures of talk' (p. 222). From a CA perspective, participation, like identity and norms, is fluid and dynamic and changes 'from syllable to syllable, from turn to turn, from action sequence to action sequence' (Sahlström 2009).

One of the challenges that these implicit norms pose for researchers is in this implicitness, which prevents them from being captured in many other approaches to analysing classroom interactions such as those used in the Learner Perspective Study (Clarke, Emanuelsson, Jablonka, & Mok 2006) or TALIS Video Study (Ingram et al. 2020). A CA approach enables us to focus on these implicit norms by focusing on patterns in teacher and student actions as they interact. In this chapter I particularly focus on how these taken for granted norms that specifically relate to what it means to do mathematics are co-constructed in interaction. Specifically, I examine the norms and practices around the use of mathematical terminology, problems and problem solving, and argumentation. These norms also construct what it means to be a mathematics student and a mathematics teacher, and what it means to do mathematics in a mathematics classroom. Before examining these norms, I first introduce the CA approach to identity.

A Conversation Analysis Approach to Identity

How do we know which attributes of a person are relevant to an interaction? A teacher may be white, female, married, have three children, play hockey at the weekends, love algebra, dread teaching geometry, and so on, but which of these attributes is relevant when analysing identities within the classroom? In Conversation Analysis it is the attributes or features that the participants themselves orient to in the interaction that matters; identity is an element of the context of interaction. This does not have to be by explicitly mentioning them. For example, the turn-taking structures in classrooms discussed in Chapter 3 mean that it is often quite clear who is the teacher and who are the students, through who controls both who speaks and what they can say. This approach to identity also means that from an ethnomethodological perspective the truth or validity of an identity category is not relevant (Wowk 2007). It is not concerned with what someone 'is', but what someone 'does'. Identities are something that is 'done' by participants in interaction; they are fluid and dynamic and co-constructed. They are also indexical to the interactional context. So, what it means to be a mathematics teacher or to be a mathematics student depends upon what actions are performed in classroom interactions. That is, what both students and teachers do when they interact. Identities considered in this way are also reflexive. By acting in a way that identifies you as a teacher, you will be making the identity of student relevant to the other participants in the interaction.

Identities are also reflexive in their relationship to both the situational context and the distal context in which the interactions occur. Zimmerman (1998) makes a distinction between discourse, situational, and transportable identities, where situational identities most closely align with Gee's (2000) notion of Discourse-identity. These are identities that are relevant to, oriented to, and co-constructed in the context in which interactions occur; for example, student and teacher. Discourse identities are more transient and connected with the sequential actions that participants in interaction perform. These include speaker, listener, questioner, answerer, and so on, and this is where the reflexive nature of identity is most visible. As one person engages in the identity of current speaker, the other participants are positioned as listeners. Transportable identities are those that carry across contexts such as gender.

The approach to identity used by both conversation analysts and discursive psychologists builds on Sack's membership categorization devices

(Sacks 1992). When we interact, we use categories to infer particular features of a situational or transportable identity. For example, the category of teacher infers features such as expertise, authority, professionalism, and so forth. These categories are not institutionally ascribed, but rather it is the demonstration of expertise and authority that places someone in the category of teacher: 'not only do categories imply features, but features imply categories' (Antaki & Widdicombe 1998, 4). It is also more than this. The meaning of expertise itself, for example, is constituted in classroom interactions.

As demonstrated in Chapters 3 and 4, the identities of teacher and student are often very visible in classroom interaction. As well as controlling turn-taking and the topic of interaction, teachers also assign tasks and the time for them, evaluate students' responses (or invite other students to make these evaluations), and tell students what to do and think. Yet what it means to be a mathematics teacher and a mathematics student varies from classroom to classroom, and person to person.

I now turn to three aspects of doing mathematics that are extensively studied in mathematics education: the communication of mathematics through using specialized terminology, problem solving, and argumentation. These three aspects are not exhaustive, but serve to illustrate what it means to do mathematics in mathematics classroom interaction, and what it means to be a mathematics student or teacher as these identities are co-constructed in interaction.

Mathematics and Language

The relationship between mathematics and language is complex, and debates about the language of mathematics, mathematics as a language, and the role of language in learning mathematics continue (Ingram, Chesnais, Erath, Rønning, & Schüler-Meyer 2020). Being able to communicate mathematics, and to communicate mathematically, is a key component of mathematics curricula around the world. Learning mathematics is far more than learning the names of concepts or processes (Morgan 2005), but these names are intrinsic to developing students' understanding of the mathematics they are learning. It is through classroom interaction that the meanings for these names are negotiated and developed. But here I focus on how students' use of some words from the specialized vocabulary of mathematics can influence the opportunities that teachers offer to use this vocabulary and the norms around the negotiation and construction of the meaning this terminology has.

An ethnomethodological approach to examining vocabulary such as CA examines the use of specialized words in interaction. These terms need to be studied in their natural habitat; within contexts that provide a need for this vocabulary, and within sentences with a communicative purpose (Moschkovich 2015). Students' use of this specialized vocabulary is reflexively related to their developing meaning for these words, and teachers provide students with opportunities to use these words that are influenced by the meanings students display.

In Extract 6-1 Teresa has been working with her students on the properties of numbers, including factors, multiples, prime, square, cube, and triangle numbers. The task involved Teresa choosing a property and the students working out what property Teresa had chosen by asking whether particular

```
133 Teresa:    hang on a second. Sam said to me, he thinks
               the number twenty-four because it's in the
               (.) [three] times table
134 Sam:           [three]
135 Sam:       and the six times table.
135 Teresa:    and the six times table,
136 Sam:       and you ((inaudible))
137 Seb:       isn't it in the eighteen times table?
138 Teresa:    would twenty-four be in the eighteen times
               table
139 Seb:       oh, no.
140 Steve:     it won't.
141 Teresa:    would-
142 Seb:       two times table
143 Teresa:    okay, okay
144 Seb:       wait (.) yeah
145 Teresa:    you are giving me (.) such good reasons. as
               well as thinking about times tables, can you
               think of the word multiples
146 Steve:     yeah.
147 Teresa:    so, could we say it's a mult- those- they are
               multiples of three, they are multiples of six
               (.) some of them are multiples of twelve,
               number twenty-four. you are so close to
               getting the answer.
148 Sophie:    it's that kind of right, because like
               (.)basically they goes in order, like six:
               like because they're going down by [threes]
149 Steve:                                         [ hey  ]
               we just said that (.) hhh
150 Teresa:    what did you just say?
151 Steve:     uh that it's in the six times table, so we
               have another number now
               ((Transcript Omitted))
190 Teresa:    what was my reason for liking those numbers?
191 Simon:     because (.) I don't know, they're special.
192 Sophie:    because, because they're all -
193 Sam:       they're multiples [of three]
```

Extract 6-1 Teresa prompting the use of multiples.

numbers between 1 and 25 had this property (http://nrich.maths.org/6962). At the point at which the extract begins, the students have identified several numbers that Teresa 'likes'.

In this example, students have been talking about numbers belonging to particular times tables, and Teresa introduces the word 'multiple' in turn 145, and models its use in turn 147. The students continue to talk about numbers belonging to times tables, as in turn 151, for some time, until turn 193. The interaction then continues, with students using the words 'multiple of' and 'times table' fluently. The students' use of the word 'multiple' is appropriate both semantically and lexically. Whilst the task is about multiples, there is no need for the students to use this word, as explaining the relationships and properties they have noticed can be effectively done through the language of 'times table'. The shift in use occurs in response to Teresa's prompting, modelling, and subsequent encouragement. At the end of the discussion Teresa closes this part of the task by saying 'it is not only because they got it right, but because they used the maths. They said they are multiples of three.' This marks the successful completion of the task as not just the identification of the property Theresa was thinking of, but also the describing of this property using the word Teresa has introduced.

This interaction includes two pedagogical foci—the identification and meaning of the relationships between the set of numbers Teresa 'likes' and naming this relationship using the word 'multiple'. The students are attending to this first focus throughout as they describe and conjecture the relationship. However, Teresa is primarily attending to the second focus and the use of the word 'multiple'. The students make conjectures around several potential relationships, some of which only apply to some of the numbers Teresa 'likes', and some of which also apply to numbers that Teresa 'does not like'. Seb, in turn 137, and Teresa, in turn 147, both imply that the relationship has to apply to all the numbers, but Teresa's evaluation focuses on the use of the word 'multiple' and the nature of the relationship; that they are *all* multiples of three is not explored explicitly. (The omitted transcripts include a third focus on the interaction between the students who are engaging in a parallel activity (Koole 2007) of who said what first.)

Later in the lesson the students are trying to identify another relationship, which is that the numbers Teresa likes are all factors of twenty. On the board, Teresa has written the numbers 5 and 10 in the box for numbers she likes, and 12 in the box for numbers she does not like.

Now in Extract 6-2 both the teacher and students are attending to the word choice, as well as the meaning of multiple. In turn 250, Sarah uses the word

```
237 Student    because it's (in the five times table).
238 Teacher    or what else could you say?
239 Student    it is a multiple -
240 Student    it's a multiple -
241 Teacher    it's a multiple of five. it's in the five
               times table, I think I can give you a stick
               for a good maths reason.
242 Student1   twenty, becau- because ten times two equals
               twenty.
243 Student    oh, no
244 Student    oh, yeah.
245 Teacher    I do, I do like twenty and your reason was?
               what was your reason for choosing twenty if
               [I'm going]
246 Student    [ oh   um  ]
247 Teacher    to give you a stick?
248 Student1   ten times two is twenty.
249 Teacher    because ten times two is twenty.
250 Sam:       and it's in the five and the ten multiples
               yeah so
251 Teacher    now,
252 Student    so
253 Teacher    you're right that twenty is ten times two.
               but I like Sam's reason. he said it was
254 Sarah:     it's a multiple of five and ten.
255 Teacher    it's a multiple of five and a multiple of
               ten.
256 Student3   and it's even
257 Student    yeah. and
258 Teacher    and it's even. okay, um, there's definitely
               some good maths talk in there.
259 Student    yeah!
260 Teacher    a lollypop stick. you haven't discovered my
               reason for my choice yet
```

Extract 6-2   Teresa prompting further use of the word 'multiple'.

'multiple' as a straight swap for the word 'times table', but note that this is not a conventional way of using the word 'multiple' in mathematics. Semantically the word is used appropriately, but not lexically. Sarah rephrases Sam's response in turn 254 in a way that is accepted by Teresa. Each of the reasons given by the students are accepted by Teresa as 'good maths talk' (turn 259), even though none of these properties or relationships describe all the numbers in the 'like' box on the board. Teresa's evaluations in turns 258 and 260 focus on the language used and whether the relationships are the one Teresa is thinking of, and not the meaning of the relationships the students have used so far.

As the task continues, Simon suggests the number 4, and Teresa includes this in the 'like' box on the board. The interaction continues in a similar way to Extract 6-2 in turn 374 Sasha makes a generalisation. Sasha is not using the language of times tables or multiples, but instead describes the relationship as 'go into' each other. Teresa picks up on this shift but focuses on the shift in

```
364 Sasha:    so they like all go into each o[ther]
365 Teresa:                                 [o::h] they're
              all going into. what is the maths word for
              the going into thing?
366 Steve:    DIVIDE
367 Teresa:   ye::s, but there's another maths word that
              was on your sheet this morning. I'm going to
              give you the stick.
368 Simon:    uh:: can we get another one if we get the
              word?
369 Teresa:   [laughs] um::
370 Simon:    uh, I think
371 Steve:    integer
372 Teresa:   you are talking about
              [maths aren't you good good]
373 Steve:    [integer,    INTEGER       ]
374 Teresa:   that wasn't the maths word, not the one that
              we're thinking of here. it's all about
              [things goi]ng into.
375 Steve:    [ subtract ]
376 Sean:     multiple.
377 Teresa:   [laughs] and it is to do with dividing, but
              it's not the word that I'm looking for.
378 Sarah:    ((inaudible))
379 Steve:    cubed num[ber ]
380 Teresa:            [what]'s the connection between all
              these?
381 Sasha:    cube number.
382 Teresa:   there's one missing. shall I put the one
              missing?
383 Simon:    triangle number.
384 Teresa:   there's one more number here that I like.
385 Sarah:    squared.
```

Extract 6-3 Teresa encourages students to use mathematical words.

language, not the shift in meaning, and asks Sasha to name the relationships. In Extract 6-3 the interaction continues with the students suggesting different names that Teresa could be after by reading out words from the worksheet in front of them. In turn 384 Teresa then tries to shift the focus back to what it is that the students are trying to name, and again in turns 390, 392, and 394. However, the students continue to attend to the activity of suggesting words. It is not until turn 461 that the word 'factor' is suggested by a student.

Again, we see this tension between the two pedagogic activities within the interaction: that of noticing and understanding the relationships between sets of numbers, and that of using mathematical terminology to name these relationships. One of the relationships within the meaning focus of the interaction is bidirectional: '4 goes into 20' means both that 20 is a multiple of 4 and that 4 is a factor of 20. There is a shift in Sasha's suggestion, from the focus on the relationships of 20 being a multiple of 4, to 4 being a factor of 20. But Teresa's prompts to think about the relationship between all the numbers and not just a pair of numbers is not taken up by the students. Students' responses to Teresa's request to name the relationship suggests that 'factor' is not yet a word that the students have active control over (and possibly integer, cube number, and triangle number too). The students are able to explain the relationship of multiples using both formal and informal language, but this is not the case for factors. Here the students have not communicated the relationship that all the numbers are factors of 20 formally or informally.

Students need opportunities to construct meaningful discourse about mathematics in order to develop their use and understanding of the language of mathematics (Schleppegrell 2007). These opportunities were adapted by Teresa throughout the lesson, depending on the meanings the students appear to have for these mathematical terms as they use them in interaction. Yet there is a tension between meaning and the use of particular words. When students had described the meaning needed using informal language, they were able to shift to using the mathematical words to communicate these mathematical ideas and relationships. When the meanings of the words were less clear, or where the concepts that the words labelled were less fully formed, the students resorted to guessing the word that Teresa was looking for.

There is also a difference in the prompts to use particular words through how Teresa generated a need for the mathematical word to be used in the interaction. The students already have a way to talk about the relationship described by the word 'multiple' using informal language, so the need for this word does not arise from the mathematics that the students are working on. Instead, it is the prompting of Teresa and the nature of the task that generates this need to use the word 'multiple'. Teresa prompts students to use the word 'multiple' in their descriptions of the relationship between the numbers on the board, to communicate their meaning. The need to generalise a relationship and to name this generalisation creates a need for the word 'factor'. Yet in the interaction, the focus is on producing this word, rather than using the word to communicate a meaning, but in this instance the students have also struggled to describe the relationship using more informal language. The students were able to identify the particular relationship of one number being a factor of another and describe this using informal language, possibly as this is the same as the relationship of one number being a multiple of another. However, generalising this to a relationship between all the numbers on the board took considerable prompting from the teacher, and this was accompanied by the challenge of identifying the word used to describe this relationship.

These interactions reveal some of the complexity of negotiating and developing meanings for relationships alongside using the generally accepted mathematical language for these relationships, even after the class has worked with a definition. There are two intertwined interactional contexts in these examples, one focused on meaning and one focused on using the target words, and whilst the teacher shifts fluidly between the two, the students do not necessarily. The relationship between these two interactional contexts is part of what it means to learn and do mathematics, negotiating the meaning of both concepts and words in communicating mathematically in interaction. This section has considered how the meaning of words is constituted in interactions between students and teachers within the mathematics classroom.

Problems and Problem Solving

Problem solving is central to learning mathematics. As Thompson (1985) states, 'to learn mathematics is to learn mathematical problem solving' (p. 190). Problem solving is also central to the OECD's definition of mathematical literacy (OECD 2018). Here I consider what problem solving is as it is constituted in interactions between teachers and students in mathematics classrooms. The examples that follow consider those tasks and activities teachers themselves describe as problems, and the actions that are performed by teachers and students around these tasks as problem solving. That is, I consider what it means to do problem solving in the mathematics classroom in interaction.

In the research literature there are a range of definitions and descriptions of 'problems' in mathematics. In mathematics we have 'word problems' that are a specific type of problem where students are asked to apply mathematical procedures to questions where words have been used to describe, disguise, or distract from the central 'problem' that is to be solved. Yet these 'word problems' are not always consistent with the generally accepted definition of a problem in mathematics as a task where there is no readily available procedure for finding the solution (Hodgen, Foster, & Kuchemann 2017, 14). Similarly, the actions involved in problem solving are widely treated as including making sense of the problem, making connections between known information, representations, and problems, as well as metacognitive activities (Guberman & Leikin 2013; Mason, Burton, & Stacey 2010; Schoenfeld 1985). These actions are similar to Hossain, Mendick, and Adler's (2013) definition of 'understanding mathematics in-depth' which includes making connections between concepts and procedures and justifying mathematical thinking, as well as explaining or communicating mathematical ideas and identifying structure and generality.

Problem solving is also often described as a cycle involving formulating, employing, and interpreting mathematics. Here, problem solving involves using mathematics to recognize the mathematical nature of a problem and formulate it in mathematical terms. This resulting mathematical problem then needs to be solved, which includes actions such as the making of strategic decisions about which mathematical processes, procedures, or facts can be used and in which order to use them. Finally, students need to evaluate the mathematical solution by interpreting the results within the original context of the problem (OECD 2018). Whilst these definitions are widely accepted in discourses about mathematics teaching, in this chapter I examine how these definitions relate to discourses within mathematics classrooms. That is, whether these definitions are oriented to by the teacher and students as they engage in problem solving in classroom interactions.

Teachers in the classrooms I observed used the word 'problem' to describe a range of tasks. For many teachers, problems are something for students to think about. In Extract 6-4, Tanya treats problems as something that can be considered differently by different students, and in Extract 6-5 Tyler asks his students to think about a question he has asked earlier as a problem.

There are also distinctions to be made between types of problem. In Extract 6-6 Tara makes a distinction and a connection between linear programming 'problems' and real-life 'problems', and later in turn 11 she also describes problems as 'sticking points'.

Each of these uses of 'problem' invites students to think and make connections, whether these connections are between the different ways that students thought about the same problem, as in Tanya's lesson, or between a game that Tyler's class have been playing and the analysis of the game using mathematical ideas and processes, or between the mathematical procedure that a class has been working on and a real life application of that procedure, as in Tara's lesson. Yet only the example from Tyler's lesson is consistent with the idea of a

```
16  Tanya:     year 9. what I would like from this remember
                (.) we're not thinking about answers but what
                I would like, as you rotate, I'm really
                interested to see if there's anything that
                makes you think about the problem
                differently. that you hadn't thought about
                okay? so just (.) as you have a look around,
                just clock that and I want to hear about it
                afterwards. is tha- is that does that make
                sense.
```

Extract 6-4 Tanya asks students to think about the problem differently.

```
193 Tyler:  … so thinking about this (1.1) as a problem.
            think about this when's it most likely to
            choose (0.4) which one? okay? what's the
            probability of the first person winning.
            what's the probability of you getting it
            right straight away.
```

Extract 6-5 Tyler asks students to think about the question as a problem.

```
6    Tara:   okay this morning we're going to be
             particularly looking at linear programming
             problems, where they are actually related to
             real life problems. so you did the mechanics
             of actually solving linear programming
             problem, I left we you with just one (.)
             actual problem type of real-life question for
             homework. so I want to see how people got on
             with that.
             ((transcript omitted))
11   Tara:   … so what I wanted to do is have a look in a
             bit of detail at that one. to sort of see how
             we solve it and where the sticking points
             were if there were, if there were problems.
             so let's start off. let's start with um (1.1)
             the, look at the question. …
```

Extract 6-6 Tara contrasts between two types of problem.

problem as something where there is no readily available procedure for finding the solution. In the example from Tanya's lesson each student has used a procedure that is readily available to them, but it may not be the same procedure as the other students. In Tara's lesson the students have been working on several similar problems that all involve the same procedures and processes, but where the numbers and the 'real life' context vary. There is therefore considerable variety in what students experience as problems in their experiences of learning mathematics. It is in the activities and actions that follow the teacher posing these problems that students develop meaning around what it means to do problem solving in their mathematics classroom. The doing of mathematics and problem solving can be very different in different classrooms.

In the three examples that follow there are some similarities in what it means to do problem solving, but there is also considerable variety in the actions that teachers and students engage in. We return first to Tara's linear programming problem, which she reads out loud in the remainder of turn 11, as shown in Extract 6-7. In the previous lessons on this topic the students have been working on formulating linear programming problems as algebraic equations, drawing the graphs of these equations, and then identifying the solution to the linear programming problem from the graphs. In this

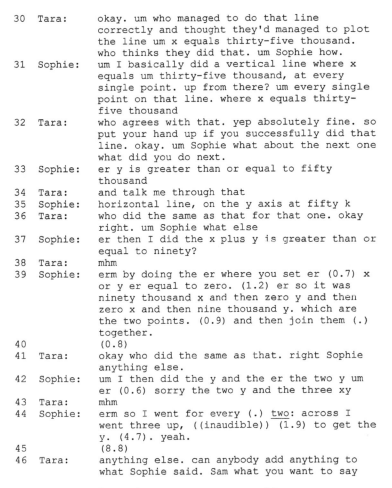

30	Tara:	okay. um who managed to do that line correctly and thought they'd managed to plot the line um x equals thirty-five thousand. who thinks they did that. um Sophie how.
31	Sophie:	um I basically did a vertical line where x equals um thirty-five thousand, at every single point. up from there? um every single point on that line. where x equals thirty-five thousand
32	Tara:	who agrees with that. yep absolutely fine. so put your hand up if you successfully did that line. okay. um Sophie what about the next one what did you do next.
33	Sophie:	er y is greater than or equal to fifty thousand
34	Tara:	and talk me through that
35	Sophie:	horizontal line, on the y axis at fifty k
36	Tara:	who did the same as that for that one. okay right. um Sophie what else
37	Sophie:	er then I did the x plus y is greater than or equal to ninety?
38	Tara:	mhm
39	Sophie:	erm by doing the er where you set er (0.7) x or y er equal to zero. (1.2) er so it was ninety thousand x and then zero y and then zero x and then nine thousand y. which are the two points. (0.9) and then join them (.) together.
40		(0.8)
41	Tara:	okay who did the same as that. right Sophie anything else.
42	Sophie:	um I then did the y and the er the two y um er (0.6) sorry the two y and the three xy
43	Tara:	mhm
44	Sophie:	erm so I went for every (.) two: across I went three up, ((inaudible)) (1.9) to get the y. (4.7). yeah.
45		(8.8)
46	Tara:	anything else. can anybody add anything to what Sophie said. Sam what you want to say

Extract 6-7   Solving a linear programming problem.

problem, the students have been given the algebraic equations in the problem and have been asked to solve the problem for homework.

Sophie describes how she drew each of the straight lines represented by the algebraic formulation of the problem given to them in turn. In turns 30 and 32 Tara treats success as drawing these lines 'correctly', but also encourages Sophie to talk through the process of drawing each of these lines. Solving the problem in this example is about following the process that has been established over the past few lessons of drawing each line in turn. Problem solving here includes making connections between an algebraic representation and a graphical representation, and identifying the solution in this graphical representation and these connections are part of the routinized process for solving problems like this. Yet the formulations, strategic decisions, and

193 Tyler:	... so thinking about this (1.1) as a problem. think about this when's it most likely to choose (0.4) which one? okay? what's the probability of the first person winning. what's the probability of you getting it right straight away.
194 Shane:	er one in ten
195 Tyler:	one in ten. good. what is the probability of the second person winning. okay, (0.5) think about it. what do we need the first person to do.
196 Sarah:	get it wrong
197 Tyler:	get it wrong. so what's the probability of the first person getting it wrong.
198 Sara:	one
199 Simon:	nine tenths
200 Tyler:	nine tenths. what's the-, an-, and we want the second person to win. so the probability of the second person winning is what.
201 Sara:	nine[ty ni]ne [((inaudible))]
202 Sid:	[eight]
203 Shane:	[sir] wha- what would you do if that [((inaudible))]
204 Seth:	what number?
205 Tyler:	it doesn't matter because we're not playing at the moment, just ((inaudible))
206 S:	oh
207 Tyler:	what's the (0.3) probability of the second person winning?
208 S:	one in nine
209 Tyler:	one in ni:ne. what do I get if I multiply those together.
210 S:	er nine[ty nine]
211 S:	[nine over ninety]
212 Tyler:	cancel it down
213	(1.4)
214 S:	three in thirty
215 S:	three in thirty
216 Tyler:	cancel it down again!
217 S:	one in ten
218 S:	one in ten
219 Tyler:	one in ten. exactly (.) the same (.) probability. second person has exactly the same chance (0.6) as the first person. the probability of the second person getting it is exactly the same. do it for the third, (0.8) we want the first person to lose. what's the probability of the second person losing. ((Transcript omitted))
	((Transcript omitted))
243 Tyler:	so I can just cancel the down straight away. so, despite what you thi:nk (0.4) it doesn't matter when you go. you still have the same (.) probability if y- if you chose before now which position to go in, you would have the same probability of winning (0.6) no matter where you go=

Extract 6-8   Tyler solving the problem with his class.

interpretation within the context of the original problem are not included in this process.

In Extract 6-8, Tyler shifts his students from playing a game involving a cup with a red cross inside it, to considering the probabilities involved in winning the game. Tyler talks about thinking about the problem several times as he introduces the idea of probability and breaks down the problem into a sequence of questions. Tyler leads the class systematically through the process of working out the probability of the first person winning and then the second person winning.

Throughout this process it is Tyler who identifies what probabilities need calculating, but it is the students who calculate the probabilities themselves. In this extract there are lots of actions that would be widely recognized as problem solving, though it is largely Tyler who engages in these actions. This process of solving the problem has not been routinized into a procedure as it was in the example from Tara's lesson in Extract 6-7, and it involves working systematically, noticing a pattern, and interpreting this pattern and the specialising of cases within the context of the original problem. Whilst Tyler has completed many of these actions associated with problem solving himself, this is in interaction with his students, and the meaning of problem solving, and what it means to do problem solving, has been co-constructed.

The Nature of Doing Mathematics

Problem solving is one example of what it means to do mathematics, yet there are a wide range of actions and behaviours involved in doing mathematics. In this section I begin by comparing and contrasting the introduction of the first task of a lesson by two of the teachers, showing how each teacher discursively constructs the nature of activity. In particular, here I focus on the teachers' choice of words within interactions as part of the action of introducing tasks. As the interaction develops, I show how the identities of mathematics teacher and student are reflexively constituted in interaction. The first three extracts come from one of Tyler's lessons on the limits of sequences, whilst the next three are from one of Tim's lessons on frequency tables.

Tyler constructs the activity in his lesson as solving problems that involve 'thinking', 'understanding', and 'having a go'. He begins by asking his students to 'look' at the problem. The problem is given using words, but Tyler has used the word 'look' rather than 'read', thus asking his students to go beyond reading the question. Tyler then describes the second part of the question as

```
18 Tyler:     okay (0.6) your fir:st thing today I've put a
              problem on the board, I will have a problem
              on the board in about (0.3) thirty seconds,
              okay I want you to look at that (.) first
              question is quite an easy one, the second
              question we have to need to think about in
              terms of (.) what it actually means, (1.3)
              okay and I want you to try your best and try
              and understand (.) how far you can get it
              done, okay. here is your problem. have a go
              at this (.) I've just inherited twelve
              thousand pounds, (0.4) okay and being the
              generous man that I am I want to donate (.)
              some of that to charity (.) but because I'm
              not totally generous, (1.2) okay I'm going to
              donate one quarter of the twelve thousand
              pounds, then the following week I want to
              donate a quarter of that amount, following
              week a quarter of that amount okay how much
              will I donate in each of the first four
              weeks, the first few are obviously easy. how
              much will you donate in total okay
```

Extract 6-9   Tyler introducing the task.

```
33 Tyler:     … wee:k two:, how much am I donating if I'm
              donating a quarter of that (.) Sam?
34 Sam:       seven point s-, seven (.) point five, no
              seven hundred and fifty
              (1.5)
35 Tyler:     ok a quarter of that, seven hundred and fifty
              pounds. ok. I want you to try and work out
              (0.8) the next (0.5) two weeks, and then I
              want you to think about (0.3) how much are
              you going to end up donating in total …
```

Extract 6-10   Tyler continues by focusing on the second part of the task.

something that 'we have to need to think about in terms of what it actually means'. By using 'we' here Tyler aligns himself with his students (Pimm 1987) in the role of problem solver. This introduction to the task refers to many of the problem-solving processes discussed above that a problem solver goes through when they encounter a problem: the processes of thinking about the problem and working out what it means, as well as what to do as the first stages of solving a problem (Mason, Burton, & Stacey 2010). In terms of the identities being constructed in this interaction, the identity of mathematics teacher includes problem posing, and the identity of mathematics student includes problem solving, thinking, trying, understanding, and having a go.

Tyler repeats Sam's answer of '750' before moving onto the next question, consequently treating this answer as appropriate and correct. Tyler also adds the units 'pounds' to the end of this answer. This echoing is usually described in the literature as a device that teachers use to encourage students to give

```
58  Tyler:    some of you used calculators, some of you
                didn't. ok that's good. I don't mind either
                way. (1.7) I want you thinking about it. ok.
                the values you got for the first three weeks
                were three thousand, (1.0) seven hundred and
                fifty, (.) one eighty seven fifty and forty
                six eighty eight?
59  Simon:    [forty-seven]
60  Tyler:    [ yep,     ] if you round it. ok (0.7) what
                I was wanting to think about is what (0.3) is
                actually happening. some of us talked about
                when do you s:top, do you stop
61  Steve:    nope
62  Tyler:    why not. hands up. (1.7) why not. Seb?
63  Seb:      because the number: (.) keeps getting
                smaller, cus of ((inaudible))
64  Tyler:    so it keeps getting smaller
65  Seb:      yep
66  Tyler:    but will there be a point where we actually
                s:top?
67  SS:       yes
68  Tyler:    why
```

Extract 6-11 Tyler asking if the sequence stops.

more complete or more mathematical answers, but it also directs attention to the specific context of the problem, that of donating money, rather than a generic calculation of a quarter of 3,000. The situational identities of teacher and student are being oriented to in this interaction through the discourse identities of questioner, evaluator, and setter of the tasks of what to do next, and reciprocally the answerers of the questions.

By mentioning calculators in turn 58, Tyler is making it relevant (in a CA sense) to the interaction. In line 59 he then reiterates that he wants his students 'thinking about it', reinforcing the earlier construction of mathematics students as thinkers. In turn 60, Tyler contrasts the performance of and the answers to the calculations summarized in turn 58 with what he wants his students to do: 'thinking'. The answers to the calculations are listed by Tyler without any description of how they were reached and without inviting the students to offer them. Whilst these calculations are a necessary part of the task, the focus is on their use in solving the problem rather than the calculations themselves. Tyler contrasts 'working it out' and 'thinking'. Performing calculations is part of the identities of both mathematics teacher and student, and so is thinking; and it is this thinking that is oriented to as something that students should do. The interaction continues with Tyler's question in turn 60 of 'do you stop'. The introduction of this question references earlier conversations ('some of us talked about' in line 65). Tyler is making these conversations relevant to the current interaction and contributes to the orientation to the question as something that is worth discussing and the answer as

```
70  Tim:       ... sometimes in the exam they won't give you
                that extra column they'll just give you these
                two, and they'll expect you to know (.) that
                it might be useful (.) to put this extra
                column on, do you know what I mean. and in a
                minute, when you do some practice from the
                text book it's the same thing. they just give
                you this bit of the table and they expect you
                to use your initiative (.) to draw in the
                extra column to do it. okay. well let's go
                through these then, the mo:de, the median,
                the mean and the range. I think we'll leave
                the mean till last because it's a bit like
                the mean one. um Sarai and (.) Sam, paying
                attention now specially, right any offers
                anyone for telling me what, m-why of course
                we always want to know why (.) what the mode,
                the median the mean and the range are. (1.7)
                ((transcript omitted)) Sophia?
71  Sophia:    er um days absent three is the mode because
                it's the most common one.
72  Tim:       right. the mode, so these are all days absent
                there's some people won't never had a day
                absent, some people have one day, some people
                had six days, some people had seven days. the
                most common number of days to have like
                absent the mode is three because a hundred
                and twenty-five people had three days off.
                that beats any o- any other sort of number of
                days off, so the mode is three. good choice
                of where to start, well done. ↑u:::m (0.4)
                >go on then< Simon
```

Extract 6-12   Tim introducing a frequency table task.

something that needs to be thought about. In this interaction, discussing is something Tyler and his students do together. This action also suggests that the answer of whether you stop or not is not immediately obvious. This is reinforced by Tyler's questions in turns 62 and 66, where he asks 'why not' twice, before re-asking the question and then asking 'why' in turn 68. This part of the problem is worth discussing and explaining. In the introduction of this problem, Tim has discursively constructed his students as students who think and find meaning, as well as students who participate in mathematical activities such as specialising, discussing, explaining, arguing, and justifying.

The next three extracts come from one of Tim's lessons and similarly illustrate how teachers can discursively construct what it means to be a mathematics student in how they introduce tasks, whilst also offering a contrast in the nature of this identity of mathematics student. This extract returns to Extract 5-29 and shows how the interaction unfolds further.

In Extract 6-12, Tim's introduction of the task includes references to doing 'some practice', needing 'to know', and using 'your initiative'. The question asked at the end of turn 70 is a 'what' question, but with a request for why

```
73  Scott:    is the ra::::nge a hundred and seven- (.)
              seventee::n
74  Tim:      range a hundred and seventeen (.) the
              ra::::nge is the ↑biggest number (.)  take
              away the smallest number (.) the ↑biggest
              number is a hundred and twenty fi:ve (0.4)
              the ↑smallest number is eight, (.) a ↑hundred
              and twenty fi:ve (.) take away ↑eight (1.4)
              Sean?
75  Sean:     no cos the, (.) the range is going to be in
              days absent so it'll be (.) eight
76  Tim:      ₀a:::h₀ re↑member Scott (0.6) this ↑table
              does ↑not have any numbers a hundred and
              twenty five in there (0.4) this ↑table only
              consists of (.) days absent from zero (0.6)
              up to ↑eight, (0.8) do you see that (1.7).
              ((transcript omitted))
89  Tim:      …um go on then (0.5) Shauna.
90  Shauna:   is the median um ta- add up all the (0.3)
              frequencies so [that] adds up to five
              hundred
```

Extract 6-13 Continuation of the introduction to the frequency table task.

included. Sophia's response is both a remembering of a definition and a procedure for finding the mode. As the interaction continues in Extract 6-13, the emphasis is on remembering and applying the procedure for calculating the range. In the introduction of this task, Tim constructs his students as tellers as well as students who do practice, use their initiative, and pay attention, and to this Sophia adds remembering definitions, applying them to questions, and using these definitions or procedures to explain answers.

This focus on remembering and applying continues, with Scott phrasing his answer in turn 73 as a question, as does Shauna in turn 90, thus orienting to the teacher's role of evaluating these answers. Tim's response to Scott also constructs the identity of teacher as also applying definitions in his description of how he saw Scott as calculating the range. The role of student as someone who remembers continues in turn 76, where Tim explicitly asks Scott to remember, but also in Shauna's turn, where she remembers the algorithm for calculating the median. There is a brief change of focus in turn 126, given in Extract 6-14.

Here Tim asks the students what these two numbers mean. Tim accepts Sheila's and Susie's answers, but his acceptance of Susie's answer returns to the procedure for calculating the mean and the role of these numbers in this calculation. In this introduction the identity of teacher has been constructed as one that involves posing questions, demonstrating, and evaluating, and the identity of student, as someone who practices, pays attention, remembers, applies, and uses definitions.

These two examples show how the identities of mathematics teacher and mathematics student can be discursively constructed in interaction, and how

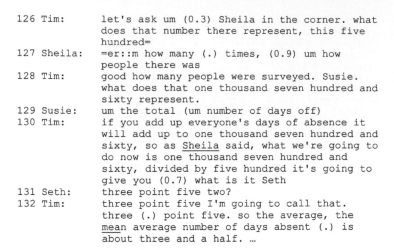

```
126 Tim:      let's ask um (0.3) Sheila in the corner. what
              does that number there represent, this five
              hundred=
127 Sheila:   =er::m how many (.) times, (0.9) um how
              people there was
128 Tim:      good how many people were surveyed. Susie.
              what does that one thousand seven hundred and
              sixty represent.
129 Susie:    um the total (um number of days off)
130 Tim:      if you add up everyone's days of absence it
              will add up to one thousand seven hundred and
              sixty, so as Sheila said, what we're going to
              do now is one thousand seven hundred and
              sixty, divided by five hundred it's going to
              give you (0.7) what is it Seth
131 Seth:     three point five two?
132 Tim:      three point five I'm going to call that.
              three (.) point five. so the average, the
              mean average number of days absent (.) is
              about three and a half. …
```

Extract 6-14 Interpreting the values in a frequency table.

these interactions constitute the nature of these identities. Teachers and students orient to and perform different mathematical activities in interaction, making particular activities relevant to these identities of mathematics teacher and student. Yet there are differences in the identities that are co-constructed. In Tyler's introduction of the task, mathematics students are discursively constructed as students who think and find meaning, who participate in mathematical activities such as specialising, discussing, explaining, arguing, and justifying. In Tim's introduction of the task, students are discursively constructed as people who practice, pay attention, remember, apply, and use definitions. What it means to be a student of mathematics as constructed through classroom interactions is specific to the context of the interaction itself, and it is not only the nature of tasks that influence the nature of this identity.

Another aspect of doing mathematics that is widely researched is proof and argumentation. These terms often encompass a range of issues, such as truth and validity, reasoning, and explanations. When considering these issues from a CA perspective, it is how the teachers and students treat them in interaction that is of interest. As Sekiguchi (1991) suggests, practices around proof, reasoning, and argumentation do not follow 'the patterns of formal mathematics'; they are shaped by the practices within the classroom. There has been extensive work on students' argumentation using ethnomethodological approaches (e.g. Knipping 2008; Krummheuer 1995; Nordin & Björklund Boistrup 2018), where arguments are co-constructed in interaction by teachers and students. Here I focus on one of the underlying purposes of argumentation and proof; that is, to persuade or convince someone of the truth or validity of a statement.

```
194 Thea:      so you want to call this a trapezium
195 Shannon:   yeah, a trapezium.  but- (1.4) with the top
               of the trapezium as well (1.2) because (1.3)
               no no (.) yeah, that's it.
196 Thea:      Have you convinced yourself it is or it
               isn't=
197 Shannon    =it is. the top right isn't a trapezium=
198 Thea:      =why not.
199 Shannon:   because both are paralle- both sides are
               parallel.
200 Thea:      both sides?  it's got four sides.
201 Shannon:   yeah, you do… it's like >top and bottom are
               parallel<, and left and right are parallel.
202 Thea:      right, we're going to talk about top and
               bottom, left and right, rather than- (.) you
               just said both sides are parallel, it doesn't
               make sense to me because there's
               fo[ur-]
203 Shannon:     [there's four] pairs.
204 Thea:      ah, right, pairs then, you're talking about
               two things=
205 Shannon:   =yeah.
206 Thea:      you've got two pairs.
207 Shannon:   yeah.
208 Thea:      and this was
209 Shannon:   one pair.
210 Thea:      you've got one pair. so, we've eliminated it
               can't be a trapezium …
```

Extract 6-15 Is a parallelogram a trapezium?

In Extract 5-32 we considered an example of a teacher (Todd) modelling the process of being convinced in mathematics. This was one of two lessons in the data collected that explicitly focused on establishing the truth of statements or conjectures. Yet issues around truth and validity can arise through other arguments and explanations. In the majority of situations, convincing is talked about as being a personal state rather than a property of a mathematical argument. Extract 6-15 is towards the end of a whole-class discussion that involves naming several quadrilaterals that are drawn on the board. The class has already discussed a rhombus as a special case of a parallelogram, and a square as a special case of a rhombus. Whilst the explicit objective of the task is to name shapes, students are also reasoning about the relationships between shapes. In turn 196, for example, the emphasis is on Shannon being convinced by her answer of 'a trapezium'.

The doubt indicated by Shannon is a particularly interesting case, as different definitions of a trapezium can result in a parallelogram being a special case of a trapezium, or a parallelogram not being a trapezium at all—and in this lesson both definitions were in use.

The doing of mathematics involves arguing, reasoning, and convincing yourself and others of the truth of a statement. What it means for a statement, relationships, or conjecture to be true is dependent upon the interactional treatment of it. In Chapter 4 I briefly explored the issue of what counts as an explanation, which also connects to the distinction made by Yackel and Cobb (1996) between norms and sociomathematical norms. However, what it means for an explanation to be mathematical is context dependent. What is accepted as an argument, as convincing, or as an explanation is also contingent upon the interactional context in which it arises. There are contexts where a 'problem is a matter of calculation, that the states in the argument we present in justification of our conclusions are the same as those we went through in getting the answer' (Krummheuer 1995, 232); that is, the distinction between explaining how and explaining why is not as clear as some researchers treat it to be.

Learning mathematics is a collaborative experience where meanings are co-constructed and negotiated in interactions between students and the teacher and between students. This learning is made possible through social, material, and interactional resources within classrooms. These resources form 'dynamic, emergent ecologies' (Abrahamson, Flood, Miele, & Siu 2019, 293) In Chapter 5 I looked at how mathematical procedures were talked about by teachers and students as things that needed to be known or remembered, and as things that are applied and used. They were not talked about as something that needs to be understood except in just two classrooms: Todd's and Tess's. Similarly, definitions were to be remembered, and there was no mention of making sense of definitions or understanding them. Chapter 4 also illustrated how particular patterns of classroom interaction implicitly give meaning to the role of errors and mistakes in mathematics teaching and learning.

The aspects of doing mathematics discussed so far are not the only features of interaction that are involved in constructing what it means to do mathematics in interaction. Another way in which what it means to do mathematics is through the use of personal pronouns. The use of 'we' is common in many mathematics classrooms (Pimm 1987). By using 'we' instead of 'you', the processes that are involved in doing mathematics or problem solving are being discussed at the general level, as something we, as a community of people who do mathematics, do. Yet there can often still be some ambiguity in who this 'we' refers to: teacher and students, teacher and other experts in mathematics, mathematics itself. The use of personal pronouns influences what it means to participate in mathematics and what it means to do mathematics. Are students doing mathematics, or doing school mathematics?

It may be that it is the nature of these tasks and activities that means that the interactions include justifications and explanations, whereas another task may mean that the focus is on remembering and applying definitions and procedures. Where one teacher, researcher, or reader may perceive contradictions between some of the practices and patterns illustrated in this chapter and their goals of teaching mathematics for understanding, these practices, patterns, and experiences are what construct for the students what it means to learn mathematics. For many students there may be no contractions or conflicts because of the meaning learning mathematics has for them.

Teachers can use a range of tasks but it is the activity around these tasks, as established through interactions between the teacher and the students, which can affect the cognitive demand of these tasks (Stein, Engle, Smith, & Hughes 2008). These interactions are also reflectively intertwined with the situational context in which they occur. These contexts reflect and are constituted by social influences, such as students' collaborative skills (Mercer & Sams 2006), motivation and engagement (Skilling et al. 2016), as well as societal influences such as the need to pass exams, and discourses of inclusion, exclusion, marginalization (Battey, Neal, Leyva, & Adams-Wiggins 2016), and ability (Lambert 2015). Learning mathematics is not a homogeneous experience, but an interactional accomplishment that is dynamic. To adapt Brouwer's and Wagner's (2004) statement about language learning, learning mathematics 'needs to be conceptualised as a social process, rather than as a social practice' (p. 32).

Conclusion

Teachers need to offer a range of ways of working mathematically in their lessons (Watson 2008). Procedures and calculations are an essential part of doing mathematics, as are generalising, justifying, and reasoning. For many educators and researchers, learning mathematics means learning to behave in mathematical ways. Burton (2004) identifies a range of mathematical activities that research mathematicians engage with that are also identified by Cuoco et al. (1996) as mathematical habits of mind. These include asking questions, making mistakes and using them, describing, explaining and discussing ideas, developing conjectures, pattern spotting, making connections between ideas and concepts, and making connections between different representations. However, research has shown that opportunities to participate in these mathematical activities is not evenly distributed across students, classrooms, or schools (Cramer & Knipping 2018; Erath 2018). Many students, particularly those with low prior attainment or from socially underprivileged backgrounds, or those who are learning mathematics in a language other than their first language, only experience and have the opportunity to engage in a very narrow range of these activities, sometimes only engaging in those activities perceived by their teacher as enabling them to pass an exam rather than to do mathematics.

The interactional work of teachers is subtle and complex and constructs what it means to do mathematics in their classrooms. Some students may only experience some mathematical activities through their teacher's modelling, whilst others may experience these through doing them for themselves, but not necessarily with an awareness of what they are doing and why it is mathematical. The examples I have offered in this chapter illustrate what it can mean to do mathematics in different mathematics classroom as teachers and students interact. The actions teachers and students perform construct different identities and meanings for mathematics students and mathematics teachers. These are not just constructed at the classroom level, but at the level of an interactional context within a lesson. Whilst many researchers define problem solving or argumentation in seemingly objective ways, these actions are in fact co-constructed through classroom interactions, and the meaning of these actions is reflexively related to the interactional context in which they occur. This has significant consequences for how students make sense of what it means to do mathematics and what it means to be a mathematics student.

7

Final Thoughts

The patterns of interaction I have described within this book have a recognizable structure that constrains and affords learning mathematics, but these patterns are also contingent upon the contexts within which they occur. Their consistency allows teachers and students agency in the variety of ways in which these patterns can be used. Characterizing and detailing these patterns of classroom interaction alongside nuancing their complexity makes them available for use by researchers, teachers, and teacher educators to make sense of what they see in the mathematics classroom.

I have raised challenges to some of the existing conceptualizations of what it means to be a successful learner of mathematics and what it means to participate successfully in learning mathematics in the classroom. As so many researchers have argued before, it is not so much the use of particular tasks or textbooks, but how these tasks are used by teachers and students that leads to successful learning. For many researchers learning is a product, whether that product is contingently produced and negotiated (e.g. Solem & Skovholt 2019), or is a concept or fact to be acquired (e.g. Purpura & Reid 2016). Studying classroom interaction often makes visible the process of learning, rather than what has been learnt. Yet learning mathematics is also about ways of reasoning, arguing, and behaving, which can also be made visible through the analysis of interaction (Eskildsen & Majlesi 2018). This book contributes to mathematics classroom research into learning in and as students' phenomena.

I also contribute to the questioning of notions of 'good teaching' or 'effective teaching'. Teacher actions are contingent upon the context in which they are performed. What might be considered 'good' teaching can not only vary between teachers, it can also vary between classrooms and within classrooms. What might be 'good' in one interactional context may not be 'good' in another. Actions and practices within classrooms change from moment to moment as teachers draw on the linguistic, social, and mathematical resources available to them. These actions and practices are also contingent upon the students' responses and interaction in the mathematics classroom.

Classrooms are heterogeneous, and teachers and students draw from a variety of linguistic, social, and multimodal resources in support of the

Patterns in Mathematics Classroom Interaction: A Conversation Analytic approach. Jenni Ingram,
Oxford University Press (2021). © Jenni Ingram. DOI: 10.1093/oso/9780198869313.003.0007

overarching pedagogical goal of an interaction. No two students, even within the same classroom, experience the same mathematics. The literature is full of descriptions of 'gaps' between different groups of learners, such as those from socially underprivileged backgrounds (e.g. Prediger 2019), or those with additional needs or disabilities (e.g. Abrahamson et al. 2019). Microanalytic approaches such as the one I have taken here can make visible how learning happens, or does not happen, in the moment-by-moment interactions that teachers and students participate in. This can illustrate the differences in the learning journeys at an individual level of observation (Douglas Fir Group 2016). What it does not do is take into account broader biographical influences or learning over a considerable time. Learning mathematics is not linear and does not follow the same trajectory for each student, even within the same classroom.

Learning includes using a variety of 'socially and culturally shaped resources for meaning making' (Bezemer 2008, 166) that are appropriated through interaction and over time. Even though many of the patterns of interaction I have described are stable across classrooms, with teachers and students orienting to these patterns as they interact, there is also considerable variation within these patterns of interaction, giving both teachers and students agency within these interactions. Other research has shown how these variations can correlate with some 'social variables' such as gender and race (Lindwall, Lymer, & Greiffenhagen 2015; Stivers & Majid 2007) and how differences in the ways students participate in classroom interactions can affect their learning (Howe et al. 2019). As Krummheuer argued, 'successful, conceptual, mathematical learning' is reflexively based on active participation (1995, 264).

Mathematics teachers mediate the mathematics that students experience. What it means to learn and do mathematics strongly depends upon a teacher's mediation (Goizueta 2019), both at the macro level, such as the research that focuses on teachers' beliefs about the nature of mathematics and mathematics learning (e.g. Beswick 2005; Goos 2013; Rebmann et al. 2015), and at the micro level (Roth & Radford 2011) that is also the focus of this book. Mathematics classroom contexts are actively constructed and maintained by both teachers and students. Student actions treated as mathematical by one teacher may not be treated as mathematical by another teacher. Teaching entails complex pedagogical decisions, judgements, and actions that respond to situational contingencies (Ehrenfeld & Horn 2020; Rowland, Huckstep, & Thwaites 2005). The patterns and structures described in this book serve as a platform for the development of strategies and solutions that respond to the challenges of mathematics learning and teaching in our ever-changing world, as well as a more nuanced understanding of existing research involving the ongoing education of teachers.

There is so much yet to discover about the process of mathematics learning as it happens in the classroom. There are also many challenges that we need to strive to overcome. Mathematics classroom interactions can open up opportunities and ways of acting that can lead to students not only behaving mathematically but also enjoying mathematics. Yet these interactions can also close down for many students what it means to do mathematics. As Hossain and colleagues argue (2013), these social practices within interaction are 'implicated in the processes of inclusion and exclusion' (p. 35), particularly when they are taken for granted or have the status of being common sense. All students draw upon a range of interactional resources, enabling them to move between interactional contexts which can also involve a shift in the nature of mathematics they are engaging in. I have seen the way that students can be labelled and marginalized in interactions, but I have also seen them empowered. Students who have been written off by many because they find mathematics difficult, in the hands of a skillful teacher have engaged in argumentation, generalising, and many other mathematical activities. Many skillful teachers support their students in shifting between the variety of interactional contexts they will find in mathematics lessons. Many of these skillful teachers have contributed data to this book.

References

Abrahamson, D., Flood, V. J., Miele, J. A., & Siu, Y.-T. (2019). 'Enactivism and ethnomethodological conversation analysis as tools for expanding Universal Design for Learning: the case of visually impaired mathematics students'. *ZDM: Mathematics Education*, 51(2), 291–303. https://doi.org/10.1007/s11858-018-0998-1.

Alexander, R. (2005). 'Culture, dialogue and learning: Notes on an emerging pedagogy'. *Education, Culture and Cognition: Intervening for Growth. International Association for Cognitive Education and Psychology (IACEP) 10th International Conference* (10–14 July), 1–4. University of Durham, UK.

Andrews, N., Ingram, J., & Pitt, A. (2016). 'The role of pauses in developing student explanations in mathematics lessons: Charlie's journey'. In G. Adams (ed.), *Proceedings of the British Society for Research into Learning Mathematics* (Vol. 36, pp. 7–12).

Antaki, C., & Widdicombe, S. (1998). 'Identity as an achievement and as a tool'. In C. Antaki & S. Widdicombe (eds), *Identities in Talk* (pp. 1–14). London: SAGE Publications Ltd.

Audacity team. (2018). *Audacity(R)*.

Barnes, R. (2007). 'Formulations and the facilitation of common agreement in meetings talk'. *Text and Talk*, 27(3), 273–96. https://doi.org/10.1515/TEXT.2007.011.

Barwell, R. (2013). 'Discursive psychology as an alternative perspective on mathematics teacher knowledge'. *ZDM: Mathematics Education*, 45(4), 595–606. https://doi.org/10.1007/s11858-013-0508-4.

Battey, D., Neal, R. A., Leyva, L., & Adams-Wiggins, K. (2016). 'The interconnectedness of relational and content dimensions of quality instruction: Supportive teacher–student relationships in urban elementary mathematics classrooms'. *The Journal of Mathematical Behavior*, 42, 1–19. https://doi.org/10.1016/j.jmathb.2016.01.001.

Bauersfeld, H. (1980). 'Hidden dimensions in the so-called reality of a mathematics classroom'. *Educational Studies in Mathematics*, 11, 23–41.

Bauersfeld, H. (1988). 'Interaction, construction, and knowledge. Alternative perspectives for mathematics education'. In D. A. Grouws & T. J. Cooney (eds), *Perspectives on Research on Effective Mathematics Teaching: Research Agenda for Mathematics Education* (pp. 27–46). Reston, VA: NCTM and Erlbaum.

BERA. (2004). *Revised Ethical Guidelines for Educational Research*. Southwell, UK: BERA.

Beswick, K. (2005). 'The beliefs/practice connection in broadly defined contexts'. *Mathematics Education Research Journal*, 17(2), 39–68. https://doi.org/10.1007/BF03217415.

Bezemer, J. (2008). 'Displaying orientation in the classroom: Students' multimodal responses to teacher instructions'. *Linguistics and Education*, 19(2), 166–78. https://doi.org/10.1016/j.linged.2008.05.005.

Bilmes, J. (1985). '"Why that now?" Two kinds of conversational meaning'. *Discourse Processes*, 8(3), 319–55. https://doi.org/10.1080/01638538509544620.

Bilmes, J. (1988). 'The concept of preference in conversation analysis'. *Language in Society*, 17(2), 161–81.

Black, P., Harrison, C., Lee, C., Marshall, B., & Wiliam, D. (2003). *Assessment for Learning: Putting it into Practice*. Maidenhead, England: Open University Press.

Blumer, H. (1969). *Symbolic Interactionism: Perspective and Method*. Englewood Cliffs, NJ: Prentice-Hall.

Boersma, P., & Weenink, D. (2018). *Praat: doing Phonetics by Computer*. http://www.praat.org/.

Brophy, J. E., & Good, T. L. (1986). 'Teacher behavior and student achievement'. In M. C. Wittrock & A. E. R. Association (eds), *Handbook of Research on Teaching* (3rd edition, pp. 328–75). New York: Macmillan.

Brousseau, G. (1997). *Theory of Didactical Situations in Mathematics* (N. Balacheff, M. Cooper, R. Sutherland, & V. Warfield, eds). Dordrecht, The Netherlands: Kluwer Academic Publishers.

Brouwer, C. E., & Wagner, J. (2004). 'Developmental issues in second language conversation'. *Journal of Applied Linguistics*, 1(1), 29–47. https://doi.org/10.1558/japl.1.1.29.55873.

Burton, L. (2004). *Mathematicians as Enquirers: Learning about Learning Mathematics*. Dordrecht, The Netherlands: Kluwer Academic Publishers.

Cameron, D. (2001). *Working with Spoken Discourse*. London: SAGE Publications Ltd.

Cazden, C. B. (2001). *Classroom Discourse: The Language of Teaching and Learning* (2nd edition). Portsmouth, NH: Heinemann.

Clarke, D., Emanuelsson, J., Jablonka, E., & Mok, I. A. C. (2006). 'The learner's perspective study and international comparisons of classsroom practice'. In D. Clarke, J. Emanuelsson, E. Jablonka, & I. A. C. Mok (eds), *Making Connections: Comparing Mathematics Classrooms around the World* (pp. 1–22). Rotterdam, The Netherlands: Sense Publishers.

Coles, A. (2013). 'Using video for professional development: The role of the discussion facilitator'. *Journal of Mathematics Teacher Education*, 16(3), 165–84. https://doi.org/10.1007/s10857-012-9225-0.

Cramer, J. C., & Knipping, C. (2018). 'Participation in argumentation'. In U. Gellert, C. Knipping, & H. Straehler-Pohl (eds), *Inside the Mathematics Class. Sociological Perspectives on Participation, Inclusion, and Enhancement* (pp. 229–44). https://doi.org/10.1007/978-3-319-79045-9_11.

Cuoco, A., Paul Goldenberg, E., & Mark, J. (1996). 'Habits of mind: An organizing principle for mathematics curricula'. *The Journal of Mathematical Behavior*, 15(4), 375–402. https://doi.org/10.1016/S0732-3123(96)90023-1.

Dörfler, W. (2016). Signs and their use: Peirce and Wittgenstein. In A. Bikner-Ahsbahs, A. Vohns, O. Schmitt, R. Bruder, & W. Dörfler (eds), *Theories in and of Mathematics Education* (pp. 21–31). https://doi.org/10.1007/978-3-319-42589-4_4.

Douglas Fir Group (2016). 'A transdisciplinary framework for SLA in a multilingual world'. *The Modern Language Journal*, 100 (Supplement 2016), 19–47.

Drew, P. (1981). 'Adults' corrections of children's mistakes: A response to Wells and Montgomery'. In P. French & M. MacLure (eds), *Adult-child Conversation* (pp. 244–67). London, UK: Croom Helm.

Drew, P., & Heritage, J. (1992). *Talk at Work: Interaction in Institutional Settings*. Cambridge, UK: Cambridge University Press.

Edwards, D., & Potter, J. (1992). 'Discursive psychology'. In M. Rapley & A. Mchoul (eds), *How to Analyse Talk in Institutional Settings: A Casebook of Methods*. London: Continuum International.

Ehrenfeld, N., & Horn, I. S. (2020). 'Initiation-entry-focus-exit and participation: A frame-work for understanding teacher groupwork monitoring routines'. *Educational Studies in Mathematics*, 103(3), 251–72.

Enfield, N. J. (2013). 'Enchrony'. In Relationship Thinking: Agency, Enchrony, and Human Sociality (pp. 583–605). https://doi.org/10.1093/acprof:oso/9780199338733.003.0004.

Enfield, N. J., & Sidnell, J. (2014). 'Language presupposes an enchronic infrastructure for social interaction'. In D. Dor, C. Knight, J. Lewis (eds), *The Social Origins of Language*, Oxford Studies in the Evolution of Language (Vol. 19) (pp. 92–104). Oxford: Oxford University Press. https://doi.org/10.1093/acprof:oso/9780199665327.003.0008.

Erath, K. (2018). 'Creating space and supporting vulnerable learners: Teachers' options for facilitating participation in oral explanations and the corresponding epistemic pro-cesses'. In R. Hunter, M. Civil, B. A. Herbel-Eisenmann, N. Planas, & D. Wagner (eds), *Mathematical Discourse that Breaks Barriers and Creates Space for Marginalized Learners* (pp. 39–60). Rotterdam, The Netherlands: Sense Publishers.

Eskildsen, S. W., & Majlesi, A. R. (2018). 'Learnables and teachables in second language talk: Advancing a social reconceptualization of central SLA tenets. Introduction to the special issue'. *The Modern Language Journal*, 102, 3–10. https://doi.org/10.1111/modl.12462.

Fasel Lauzon, V., & Berger, E. (2015). 'The multimodal organization of speaker selection in classroom interaction'. *Linguistics and Education*, 31, 14–29. https://doi.org/10.1016/j.linged.2015.05.001.

Forman, E. A., Larreamendy-Joerns, J., Stein, M. K., & Brown, C. A. (1998). '"You're going to want to find out which and prove it": Collective argumentation in a mathematics classroom'. *Learning and Instruction*, 8(6), 527–48. https://doi.org/10.1016/S0959-4752(98)00033-4.

Franke, M. L., Webb, N. M., Chan, A., Battey, D., Ing, M., Freund, D., & De, T. (2007). 'Eliciting student thinking in elementary school mathematics classrooms'. In *CRESST Report 725*.

Franke, M. L., Webb, N. M., Chan, A. G., Ing, M., Freund, D., & Battey, D. (2009). 'Teacher questioning to elicit students' mathematical thinking in elementary school classrooms'. *Journal of Teacher Education*, 60(4), 380–92. https://doi.org/10.1177/0022487109339906.

Furtak, E. M., & Shavelson, R. J. (2009). 'Guidance, conceptual understanding and student learning: An investigation of inquiry-base teaching in the US'. In *The Power of Video Studies in Investigating Teaching and Learning in the Classroom* (pp. 181–206). Münster: Waxman.

Gardner, R. (2019). 'Classroom interaction research: The state of the art'. *Research on Language and Social Interaction*, 52(3), 212–26. https://doi.org/10.1080/08351813.2019.1631037.

Garfinkel, H. (1967). *Studies in Ethnomethodology*. Englewood Cliffs, NJ: Prentice-Hall.

Gee, J. P. (2000). 'Identity as an analytic lens for research in education'. *Review of Research in Education*, 25(1), 99–125. https://doi.org/10.3102/0091732X025001099.

Goffman, E. (1976). 'Replies and responses'. *Language in Society*, 5(3), 257–313.

Goffman, E. (1981). *Forms of Talk*. Oxford: Blackwell.

Goizueta, M. (2019). 'Epistemic issues in classroom mathematical activity: There is more to students' conversations than meets the teacher's ear'. *The Journal of Mathematical Behavior*, 55, 100691. https://doi.org/10.1016/j.jmathb.2019.01.007.

Goodwin, C., & Goodwin, M. H. (2004). 'Participation'. In A. Duranti (ed.), *A Companion to Linguistic Anthropology* (pp. 222–44). Oxford: Blackwell Publishers Ltd.

Goodwin, M. H. (1999). 'Participation'. *Journal of Linguistic Anthropology*, 9(1–2), 177–80.

Goos, M. (2013). 'Sociocultural perspectives in research on and with mathematics teachers: A zone theory approach'. *ZDM: Mathematics Education*, 45(4), 521–33. https://doi.org/10.1007/s11858-012-0477-z.

Greenleaf, C., & Freedman, S. W. (1993). 'Linking classroom discourse and classroom content: Following the trail of intellectual work in a writing lesson'. *Discourse Processes*, 16(4), 465–505. https://doi.org/10.1080/01638539309544850.

Guberman, R., & Leikin, R. (2013). 'Interesting and difficult mathematical problems: changing teachers' views by employing multiple-solution tasks'. *Journal of Mathematics Teacher Education*, 16(1), 33–56. https://doi.org/10.1007/s10857-012-9210-7.

Hall, J. K. (2018). 'From L2 interactional competence to L2 interactional repertoires: reconceptualising the objects of L2 learning'. *Classroom Discourse*, 9(1), 25–39. https://doi.org/10.1080/19463014.2018.1433050.

Harré, R., & Van Langenhove, L. (1999). *Positioning Theory: Moral Contexts of Intentional Action*. Oxford: Blackwell.

Heinemann, T., Lindström, A., & Steensig, J. (2011). 'Addressing epistemic incongruence in question–answer sequences through the use of epistemic adverbs'. In T. Stivers, L. Mondada, & J. Steensig (eds), The Morality of Knowledge in Conversation (pp. 107–30). Cambridge, UK: Cambridge University Press. https://doi.org/10.1017/CBO9780511921674.006.

Hepburn, A., & Bolden, G. B. (2013). 'The Conversation Analytic approach to transcription'. In J. Sidnell & T. Stivers (eds), *The Handbook of Conversation Analysis* (pp. 57–76). https://doi.org/10.1002/9781118325001.ch4.

Herbel-Eisenmann, B. A. A. (2000). *How Discourse Structures Norms: A Tale of Two Middle School Mathematics Classrooms* (Michigan State University). https://doi.org/10.16953/deusbed.74839

Herbst, P., & Kosko, K. W. (2013). 'Using representations of practice to elicit mathematics teachers' tacit knowledge of practice: A comparison of responses to animations and videos'. *Journal of Mathematics Teacher Education*, 17(6), 515–37. https://doi.org/10.1007/s10857-013-9267-y.

Heritage, J. (1984). *Garfinkel and Ethnomethodology*. New York: Polity Press.

Heritage, J. (2012a). 'Epistemics in action: Action formation and territories of knowledge'. *Research on Language and Social Interaction*, 45(1), 1–29. https://doi.org/10.1080/08351813.2012.646684.

Heritage, J. (2012b). 'The epistemic engine: Sequence organization and territories of knowledge'. *Research on Language and Social Interaction*, 45(1), 30–52. https://doi.org/10.1080/08351813.2012.646685.

Heritage, J. (2013). 'Epistemics in conversation'. In J. Sidnell & T. Stivers (eds), *The Handbook of Conversation Analysis* (pp. 370–94). https://doi.org/10.1002/9781118325001.ch18.

Heritage, J., & Raymond, G. (2005). 'The terms of agreement: indexing epistemic authority and subordination in talk-in-interaction'. *Social Psychology Quarterly*, 68(1), 15–38. https://doi.org/10.1177/019027250506800103.

Heritage, J., & Watson, R. (1979). 'Formulations as conversational objects'. *Everyday Language*, (November), 123–62.

Hodgen, J., Foster, C., & Kuchemann, D. (2017). *Improving Mathematics in Key Stages Two and Three*. Retrieved from https://educationendowmentfoundation.org.uk/public/files/Publications/Campaigns/Maths/KS2_KS3_Maths_Guidance_2017.pdf.

Hofmann, R., & Ruthven, K. (2018). 'Operational, interpersonal, discussional and idea-tional dimensions of classroom norms for dialogic practice in school mathematics'. *British Educational Research Journal*, 44(3), 496–514. https://doi.org/10.1002/berj.3444

Hogan, D., Rahim, R. A., Chan, M., Kwek, D., & Towndrow, P. (2012). 'Understanding classroom talk in secondary three mathematics classes in Singapore'. In B. Kaur & T. L. Toh (eds), *Reasoning, Communication and Connections in Mathematics: Yearbook 2012, Association of Mathematics Educators* (pp. 169–97). https://doi.org/ 10.1142/9789814405430_0009.

Hossain, S., Mendick, H., & Adler, J. (2013). 'Troubling "understanding mathematics in-depth": Its role in the identity work of student-teachers in England'. *Educational Studies in Mathematics*, 84(1), 35–48. https://doi.org/10.1007/s10649-013-9474-6.

Howe, C., Hennessy, S., Mercer, N., Vrikki, M., & Wheatley, L. (2019). 'Teacher–student dialogue during classroom teaching: Does it really impact on student outcomes?' *Journal of the Learning Sciences*, 28(4–5), 462–512. https://doi.org/10.1080/1050840 6.2019.1573730.

Hutchby, I., & Wooffitt, R. (1998). *Conversation Analysis: Principles, Practices and Applications*. Cambridge: Polity.

Ingram, J. (2010). 'The affordances and constraints of turn-taking in the secondary math-ematics classroom'. In M. Joubert (ed.), *Proceedings of the British Society for Research into Learning Mathematics* (pp. 49–54). Retrieved from https://bsrlm.org.uk/wp-content/uploads/2016/02/BSRLM-IP-30-3-09.pdf.

Ingram, J. (2012). 'Whole class interaction in the mathematics classroom: A conversation analytic approach'. PhD Thesis, University of Warwick.

Ingram, J. (2014). 'Shifting attention'. *For the Learning of Mathematics*, 34(3), 19–24.

Ingram, J. (2018). 'Moving forward with ethnomethodological approaches to analysing mathematics classroom interactions'. *ZDM¬¬: Mathematics Education*, 50(6), 1065–75. https://doi.org/10.1007/s11858-018-0951-3.

Ingram, J., & Andrews, N. (2019). 'Claims and demonstrations of understanding in whole class interactions'. *Proceedings of the 11th Congress of Research in Mathematics Education, February 6th–10th*. Utrecht, The Netherlands.

Ingram, J., Andrews, N., & Pitt, A. (2016). 'Patterns of interaction that encourage student explanations in mathematics lessons'. In G. Adams (ed.), *Proceedings of the British Society for Research into Learning Mathematics* (pp. 37–41).

Ingram, J., Andrews, N., & Pitt, A. (2017). 'Revisiting the roles of interactional patterns in mathematics classroom interaction'. In T. Dooley & G. Gueudet (eds), *Proceedings of the 10th Congress of Research in Mathematics Education* (pp. 1300–7). Dublin, Ireland: DCU and ERME.

Ingram, J., Andrews, N., & Pitt, A. (2019). 'When students offer explanations without the teacher explicitly asking them to'. *Educational Studies in Mathematics*, 101(1), 51–66. https://doi.org/10.1007/s10649-018-9873-9.

Ingram, J., Chesnais, A., Erath, K., Rønning, F., & Schüler-Meyer, A. (2020). 'Language in the mathematics classroom'. In J. Ingram, K. Erath, F. Rønning, & A. Schüler-Meyer (eds), *Proceedings of the Seventh ERME Topic Conference* (pp. 1–8). Montpellier: ERME.

Ingram, J., & Elliott, V. (2014). 'Turn taking and "wait time" in classroom interactions'. *Journal of Pragmatics*, 62, 1–12. https://doi.org/10.1016/j.pragma.2013.12.002.

Ingram, J., & Elliott, V. (2016). 'A critical analysis of the role of wait time in classroom interactions and the effects on student and teacher interactional behaviours'. *Cambridge Journal of Education*, 46(1), 1–17. https://doi.org/10.1080/0305764X.2015.1009365.

Ingram, J., & Elliott, V. (2019). *Research Methods for Classroom Discourse*. London: Bloomsbury.

Ingram, J., A. Lindorff, P. Sammons, T. McDermott, P. Mitchell, K. Smith, & M. Voss (2020). *TALIS Video Study: England National Report*. London: DfE.

Ingram, J., Pitt, A., & Baldry, F. (2015). 'Handling errors as they arise in whole-class interactions'. *Research in Mathematics Education*, 17(3), 183–97. https://doi.org/10.108 0/14794802.2015.1098562.

Jefferson, G. (1989). 'Preliminary notes on a possible metric which provides for a "standard maximum" silence of approximately one second in conversation'. In D. Roger & P. Bull (eds), *Conversation: An Interdisciplinary Perspective* (pp. 156–97). Clevedon: Multilingual Matters.

Jefferson, G. (2004). 'Glossary of transcript symbols with an introduction'. In G. H. Lerner (ed.), *Conversation Analysis: Studies from the First Generation* (pp. 13–31). Amsterdam/ Philadelphia: John Benjamins Publishing Company.

Ju, M. K., & Kwon, O. N. (2007). 'Ways of talking and ways of positioning: Students' beliefs in an inquiry-oriented differential equations class'. *The Journal of Mathematical Behavior*, 26(3), 267–80. https://doi.org/10.1016/j.jmathb.2007.10.002.

Käntä, L. (2012). 'Teachers' embodied allocations in instructional interaction'. *Classroom Discourse*, 3(2), 166–86. https://doi.org/10.1080/19463014.2012.716624.

Kapellidi, C. (2015). 'The interplay between agency and constraint: Some departures from the organization of talk in the classroom'. *Text and Talk*, 35(4), 453–79. https://doi. org/10.1515/text-2015-0012.

Knipping, C. (2008). 'A method for revealing structures of argumentations in classroom proving processes'. *ZDM: Mathematics Education*, 40(3), 427–41. https://doi. org/10.1007/s11858-008-0095-y.

Koole, T. (2007). 'Parallel activities in the classroom'. *Language and Education*, 21(6), 487–501. https://doi.org/10.2167/le713.0.

Koole, T. (2010). 'Displays of epistemic access: Student responses to teacher explanations'. *Research on Language and Social Interaction*, 43(2), 183–209. https://doi. org/10.1080/08351811003737846.

Koole, T., & Berenst, J. (2008). 'Pupil participation in plenary interaction'. In J. Deen, M. Hajer, & T. Koole (eds), *Interaction in Two Multicultural Mathematics Classrooms* (pp. 107–38). Amsterdam: Aksant.

Koshik, I. (2005). 'Alternative questions used in conversational repair'. *Discourse Studies*, 7(2), 193–211. https://doi.org/10.1177/1461445605050366.

Krummheuer, G. (1995). 'The ethnography of argumentation'. In P. Cobb & H. Bauersfeld (eds), *The Emergence of Mathematical Meaning. Interaction in Classroom Cultures* (pp. 229–70). Hillsdale, NJ: Lawrence Erlbaum Associates Inc.

Krummheuer, G. (2007). 'Argumentation and participation in the primary mathematics classroom. Two episodes and related theoretical abductions'. *The Journal of Mathematical Behavior*, 26(1), 60–82. https://doi.org/10.1016/j.jmathb.2007.02.001.

Krummheuer, G. (2015). 'Methods for reconstructing processes of argumentation and participation in primary mathematics classroom interaction'. In A. Bikner-Ahsbahs, C. Knipping, & N. Presmeg (eds), *Approaches to Qualitative Research in Mathematics Education: Examples of Methodology and Methods* (pp. 51–74). Dordrecht, The Netherlands: Springer. https://doi.org/10.1007/978-94-017-9181-6.

Kyriacou, C., & Issitt, J. (2008). What Characterises Effective Teacher-initiated Teacher-pupil Dialogue to Promote Conceptual Understanding in Mathematics Lessons in

England in Key Stages 2 and 3? London: EPPI-Centre, Social Research Unit, Institute of Education.

Lambert, R. (2015). 'Constructing and resisting disability in mathematics classrooms: a case study exploring the impact of different pedagogies'. *Educational Studies in Mathematics*, 89(1), 1–18. https://doi.org/10.1007/s10649-014-9587-6.

Lave, J., & Wenger, E. (1991). *Situated Learning: Legitimate Peripheral Participating*. Cambridge, UK: Cambridge University Press.

Lawler, B. R. (2018). 'Learning to support student discourse in an urban high school district'. In R. Hunter, M. Civil, B. A. Herbel-Eisenmann, N. Planas, & D. Wagner (eds), *Mathematics Discourse that Breaks Barriers and Creates Space for Marginalized Learners* (pp. 121–46). Rotterdam, The Netherlands: Sense Publishers.

Lee, Y.-A. (2007). 'Third turn position in teacher talk: Contingency and the work of teaching'. *Journal of Pragmatics*, 39(6), 1204–30. https://doi.org/10.1016/j.pragma.2006.11.003.

Lee, Y.-A. (2010). 'Learning in the contingency of talk-in-interaction'. *Text and Talk*, 30(4), 403–22. https://doi.org/10.1515/TEXT.2010.020.

Leinhardt, G. (2001). 'Instructional explanations: A commonplace for teaching and location for contrast'. In V. Richardson (cd.), *Handbook of Research on Teaching* (4th edition, pp. 333–57). Washington, DC: American Educational Research Association.

Lerner, G. H. (2002). 'Turn-sharing: The choral co-production of talk-in-interaction'. *The Language of Turn and Sequence*, 225–56. Retrieved from http://www.soc.ucsb.edu/faculty/lerner/pub/Turn_Sharing.pdf.

Levenson, E., Tirosh, D., & Tsamir, P. (2009). 'Students' perceived sociomathematical norms: The missing paradigm'. *The Journal of Mathematical Behavior*, 28(2–3), 171–87. https://doi.org/10.1016/j.jmathb.2009.09.001.

Liebscher, G., & Dailey-O'Cain, J. (2003). 'Conversational repair as a role-defining mechanism in classroom interaction'. *The Modern Language Journal*, 87(3), 375–90. https://doi.org/10.1111/1540-4781.00196.

Lindwall, O., & Lymer, G. (2011). 'Uses of "understand" in science education'. *Journal of Pragmatics*, 43(2), 452–74. https://doi.org/10.1016/j.pragma.2010.08.021.

Lindwall, O., Lymer, G., & Greiffenhagen, C. (2015). 'The sequential analysis of instruction'. In N. Markee (ed.), *The Handbook of Classroom Discourse and Interaction* (pp. 142–57), Oxford: Wiley-Blackwell. https://doi.org/10.1002/9781118531242.ch9

Macbeth, D. (2001). 'On "reflexivity" in qualitative research: Two readings, and a third'. *Qualitative Inquiry*, 7(1), 35–68. https://doi.org/10.1177/107780040100700103.

Macbeth, D. (2003). 'Hugh Mehan's *Learning Lessons* reconsidered: On the differences between the naturalistic and critical analysis of classroom discourse'. *American Educational Research Journal*, 40(1), 239–80. https://doi.org/10.3102/00028312040001239.

Macbeth, D. (2004). 'The relevance of repair for classroom correction'. *Language in Society*, 33(5), 703–36. https://doi.org/10.1017/S0047404504045038.

Macbeth, D. (2011). 'Understanding understanding as an instructional matter'. *Journal of Pragmatics*, 43(2), 438–51. https://doi.org/10.1016/j.pragma.2008.12.006.

Margutti, P., & Drew, P. (2014). 'Positive evaluation of student answers in classroom instruction'. *Language and Education*, 28(5), 436–58. https://doi.org/10.1080/09500782.2014.898650.

Mason, J. (2002). *Researching your Own Practice: The Discipline of Noticing*. Abingdon, Oxford: Routledge.

Mason, J. (2012). 'Noticing: Roots and branches'. In M. G. Sherin, V. R. Jacobs, & R. A. Philipp (eds), *Mathematics Teacher Noticing: Seeing Through Teachers' Eyes* (pp. 35–50). New York: Routledge.

Mason, J. (2016). 'When is a problem…? "when" is actually the problem!' In P. Felmer, E. Pehkonen, & J. Kilpatrick (eds), *Posing and Solving Mathematical Problems: Advances and New Perspectives* (pp. 263–85). https://doi.org/10.1007/978-3-319-28023-3.

Mason, J., Burton, L., & Stacey, K. (2010). *Thinking Mathematically* (2nd edition). London: Pearson. https://doi.org/10.12968/eyed.2013.15.2.18

McHoul, A. (1978). 'The organization of turns at formal talk in the classroom'. *Language in Society*, 7(2), 183–213.

McHoul, A. (1990). 'The organization of repair in classroom talk'. *Language in Society*, 19(3), 349–77.

Mehan, H. (1979a). *Learning Lessons: Social Organization in the Classroom*. Cambridge, MA: Harvard University Press.

Mehan, H. (1979b). '"What time is it, Denise?": Asking known information questions in classroom discourse'. *Theory into Practice*, 18(4), 285–94.

Mercer, N., & Sams, C. (2006). 'Teaching children how to use language to solve maths problems'. *Language and Education*, 20(6), 507–28. https://doi.org/10.2167/le678.0.

Michaels, S., & O'Connor, C. (2015). 'Conceptualizing talk moves as tools: Professional development approaches for academically productive discussions'. In L. Resnick, C. Asterhan, & S. Clarke (eds), *Socializing Intelligence Through Academic Talk and Dialogue* (pp. 347–61). Washington, DC: American Educational Research Association.

Morgan, C. (2005). 'Word, definitions and concepts in discourses of mathematics, teaching and learning'. *Language and Education*, 19(2), 102–16. https://doi.org/10.1080/09500780508668666.

Moschkovich, J. N. (2015). 'Academic literacy in mathematics for English Learners'. *The Journal of Mathematical Behavior*, 40, 43–62. https://doi.org/10.1016/j.jmathb.2015.01.005

Moyer, P. S., & Milewicz, E. (2002). 'Learning to question: Categories of questioning used by preservice teachers during diagnostic mathematics interviews'. *Journal of Mathematics Teacher Education*, 5, 293–315.

Nassaji, H., & Wells, G. (2000). 'What's the use of "triadic dialogue"?: An investigation of teacher-student interaction'. *Applied Linguistics*, 21(3), 376–406.

Nolan, K. (2012). 'Dispositions in the field: Viewing mathematics teacher education through the lens of Bourdieu's social field theory'. *Educational Studies in Mathematics*, 80(1–2), 201–15. https://doi.org/10.1007/s10649-011-9355-9.

Nordin, A. K., & Björklund Boistrup, L. (2018). 'A framework for identifying mathematical arguments as supported claims created in day-to-day classroom interactions'. *The Journal of Mathematical Behavior*, 51, 15–27. https://doi.org/10.1016/j.jmathb.2018.06.005.

O'Connor, C., & Michaels, S. (1993). 'Aligning academic task and participation status through revoicing: Analysis of a classroom discourse strategy'. *Anthropology and Education Quarterly*, 24(4), 318–35.

O'Connor, C., & Michaels, S. (1996). 'Shifting participant frameworks: Orchestrating thinking practices in group discussion'. In D. Hicks (ed.), *Discourse, Learning, and Schooling* (pp. 63–103). Cambridge, UK: Cambridge University Press. https://doi.org/10.1017/cbo9780511720390.003.

O'Connor, C., & Michaels, S. (2019). 'Supporting teachers in taking up productive talk moves: The long road to professional learning at scale'. *International Journal of Educational Research*, 97, 166–75. https://doi.org/10.1016/j.ijer.2017.11.003.

OECD. (2018). *PISA 2021 Mathematics Framework*. Paris.

Payne, G., & Hustler, D. (1980). 'Teaching the class: The practical management of a cohort'. *British Journal of Sociology of Education*, 1(1), 49–66. https://doi.org/10.1080/0142569800010104.

Pimm, D. (1987). *Speaking Mathematically: Communication in Mathematics Classrooms*. London: Routledge and Kegan Paul.

Planas, N., & Gorgorió, N. (2004). 'Are different students expected to learn norms differently in the mathematics classroom?' *Mathematics Education Research Journal*, 16(1), 19–40. https://doi.org/10.1007/BF03217389.

Pomerantz, A. (1984). 'Agreeing and disagreeing with assessments: Some features of preferred/dispreferred shapes'. In J. Atkinson & J. Heritage (eds), *Structures of Social Action: Studies in Conversation Analysis*. New York: Cambridge University Press.

Pomerantz, A., & Heritage, J. (2013). 'Preference'. In J. Sidnell & T. Stivers (eds), *The Handbook of Conversation Analysis* (pp. 210–28). Chichester, UK: Wiley-Blackwell.

Prediger, S. (2019). 'Investigating and promoting teachers' expertise for language-responsive mathematics teaching'. *Mathematics Education Research Journal*, 31(4), 367–92. https://doi.org/10.1007/s13394-019-00258-1.

Purpura, D. J., & Reid, E. E. (2016). 'Mathematics and language: Individual and group differences in mathematical language skills in young children'. *Early Childhood Research Quarterly*, 36, 259–68. https://doi.org/10.1016/j.ecresq.2015.12.020.

Rebmann, K., Schloemer, T., Berding, F., Luttenberger, S., & Paechter, M. (2015). 'Preservice teachers' personal epistemic beliefs and the beliefs they assume their pupils to have'. *European Journal of Teacher Education*, 38(3), 284–99. https://doi.org/10.1080/02619768.2014.994059.

Roth, W.-M., & Gardner, R. (2012). '"They're gonna explain to us what makes a cube a cube?" Geometrical properties as contingent achievement of sequentially ordered child-centered mathematics lessons'. *Mathematics Education Research Journal*, 24, 323–46.

Roth, W.-M., & Radford, L. (2011). *A Cultural-historical Perspective on Mathematics and Learning*. Rotterdam, The Netherlands: Sense Publishers.

Rowe, M. B. (1972). 'Wait-time and rewards as instructional variables: Their influence on language, logic and fate control'. *National Association for Research in Science Teaching*. Chicago, Illinois.

Rowe, M. B. (1986). 'Wait time: Slowing down may be a way of speeding up!' *Journal of Teacher Education*, 37(1), 43–50. https://doi.org/10.1177/002248718603700110.

Rowland, T. (1995). 'Hedges in mathematics talk: Linguistic pointers to uncertainty'. *Educational Studies in Mathematics*, 29, 327–53.

Rowland, T., Huckstep, P., & Thwaites, A. (2005). 'Elementary teachers' mathematics subject knowledge: The knowledge quartet and the case of Naomi'. *Journal of Mathematics Teacher Education*, 8(3), 255–81. https://doi.org/10.1007/s10857-005-0853-5.

Sacks, H. (1992). *Lectures on Conversation* (G. Jefferson, ed.). Oxford, UK: Blackwell.

Sacks, H., Schegloff, E. A., & Jefferson, G. (1974). 'A simplest systematics for the organization of turn-taking for conversation'. *Language*, 50(4), 696–735.

Sahlström, F. (2009). 'Editorial'. *Scandinavian Journal of Educational Research*, 53(2), 103–11. https://doi.org/10.1080/00313830902757543.

Schegloff, E. A. (2000). 'Overlapping talk and the organization of turn-taking for conversation'. *Language in Society*, 29, 1–63.

Schegloff, E. A. (2007). *Sequence Organization in Interaction: A Primer in Conversation Analysis*. Cambridge, UK: Cambridge University Press.

Schegloff, E. A., Jefferson, G., & Sacks, H. (1977). 'The preference for self-correction in the organization of repair in conversation'. *Language*, 53(2), 361–82.

Schegloff, E. A., & Sacks, H. (1973). 'Opening up closings'. *Semiotica*, 8(4), 289–327. https://doi.org/10.1515/semi.1973.8.4.289.

Schleppegrell, M. (2007). 'The linguistic challenges of mathematics teaching and learning: A research review'. *Reading and Writing Quarterly*, 23(2), 139–59. https://doi.org/10.1080/10573560601158461.

Schoenfeld, A. H. (1985). *Mathematical Problem Solving*. Orlando, FL: Academic Press.

Schoenfeld, A. H. (2018). 'Video analyses for research and professional development: The teaching for robust understanding (TRU) framework'. *ZDM: Mathematics Education*, 50(3), 491–506. https://doi.org/10.1007/s11858-017-0908-y.

Seedhouse, P. (1996). *Learning Talk: A Study of the Interactional Organisation of the L2 Classroom from a CA Institutional Discourse Perspective*. York: University of York.

Seedhouse, P. (1997). 'The case of the missing "No": The relationship between pedagogy and interaction'. *Language Learning*, 47(3), 547–83.

Seedhouse, P. (2004). *The Interactional Architecture of the Language Classroom: A Conversation Analysis Perspective*. Malden, MA: Blackwell.

Seedhouse, P. (2019). 'L2 classroom contexts: deviance, confusion, grappling and flouting'. *Classroom Discourse*, 10(1), 10–28. https://doi.org/10.1080/19463014.2018.1555768.

Sekiguchi, Y. (1991). 'An investigation on proofs and refutations in the mathematics classroom'. Ed.D. Dissertation, University of Georgia.

Sekiguchi, Y. (2006). 'Mathematical norms in Japanese mathematics lessons'. In D. Clarke, C. Keitel, & Y. Shimizu (eds), *Mathematics Classrooms in Twelve Countries: The Insider's Perspective* (pp. 289–306). Rotterdam, The Netherlands: Sense Publishers.

Sidnell, J. (2010). *Conversation Analysis: An Introduction*. Chichester, England: Wiley-Blackwell.

Sidnell, J., & Stivers, T. (2012). *The Handbook of Conversation Analysis*. Oxford: Wiley-Blackwell.

Sinclair, J. M. H., & Coulthard, M. (1975). *Towards an Analysis of Discourse: The English Used by Teachers and Pupils*. London: Oxford University Press.

Skilling, K., Bobis, J., Martin, A. J., Anderson, J., & Way, J. (2016). 'What secondary teachers think and do about student engagement in mathematics'. *Mathematics Education Research Journal*, 28(4), 545–66. https://doi.org/10.1007/s13394-016-0179-x.

Smith, H., & Higgins, S. (2006). 'Opening classroom interaction: The importance of feedback'. *Cambridge Journal of Education*, 36(4), 485–502. https://doi.org/10.1080/03057640601048357.

Solem, M. S., & Skovholt, K. (2019). 'Teacher formulations in classroom interactions'. *Scandinavian Journal of Educational Research*, 63(1), 69–88.

Stein, M. K., Engle, R. A., Smith, M. S., & Hughes, E. K. (2008). 'Orchestrating productive mathematical discussions: Five practices for helping teachers move beyond show and tell'. *Mathematical Thinking and Learning*, 10(4), 313–40. https://doi.org/10.1080/10986060802229675.

Stivers, T. (2008). 'Stance, alignment, and affiliation during storytelling: When nodding is a token of affiliation'. *Research on Language and Social Interaction*, 41(1), 31–57. https://doi.org/10.1080/08351810701691123.

Stivers, T., & Majid, A. (2007). 'Questioning children: Interactional evidence of implicit bias in medical interviews'. *Social Psychology Quarterly*, 70(4), 424–41. https://doi.org/10.1177/019027250707000410.

Stivers, T., & Robinson, J. D. (2006). 'A preference for progressivity in interaction'. *Language in Society*, 35(3), 367–92. https://doi.org/10.1017/S0047404506060179.

Temple, C., & Doerr, H. M. (2012). 'Developing fluency in the mathematical register through conversation in a tenth-grade classroom'. *Educational Studies in Mathematics*, 81(3), 287–306. https://doi.org/10.1007/s10649-012-9398-6.

ten Have, P. (2007). *Doing Conversation Analysis* (2nd edition). London, UK: SAGE Publications Ltd.

Thompson, P. W. (1985). 'Experience, problem solving, and learning mathematics: Considerations in developing mathematics curricula'. In E. A. Silver (ed.), *Teaching and Learning Mathematical Problem Solving: Multiple Research Perspectives* (pp. 189–243). Hillsdale, NJ: Erlbaum.

Tobin, K. (1986). 'Effects of teacher wait time on discourse characteristics in mathematics and language arts classes'. *American Educational Research Journal*, 23(2), 191–200.

Tsui, A. B. M. (1991). 'Sequencing rules and coherence in discourse'. *Journal of Pragmatics*, 15(2), 111–29.

van Lier, L. (1988). *The Classroom and the Language Learner*. London: Longman.

Venkat, H., Askew, M., Watson, A., Mason, J., Venkatakrishnan, H., Askew, M., . . . Mason, J. (2019). 'Architecture of mathematical structure'. *For the Learning of Mathematics*, 39(1), 13–17.

Vogler, A., Prediger, S., Quasthoff, U., & Heller, V. (2018). 'Students' and teachers' focus of attention in classroom interaction: Subtle sources for the reproduction of social disparities'. *Mathematics Education Research Journal*, 30(4), 299–323.

Walsh, S. (2011). *Exploring Classroom Discourse: Language in Action*. Abingdon; New York: Routledge.

Waring, H. Z. (2009). 'Moving out of IRF (Initiation-Response-Feedback): A single case analysis'. *Language Learning*, 59(4), 796–824.

Waring, H. Z. (2014). 'Turn-allocation and context: Broadening participation in the second language classroom'. In J. Flowerdew (ed.), *Discourse in Context: Contemporary Applied Linguistics* (Vol. 3, pp. 301–20). https://doi.org/10.5040/9781474295345.0018.

Watson, A. (2008). 'School mathematics as a special kind of mathematics'. *For the Learning of Mathematics*, 28(3), 3–7.

Weatherall, A., & Keevallik, L. (2016). 'When claims of understanding are less than affiliative'. *Research on Language and Social Interaction*, 49(3), 167–82. https://doi.org/10.1080/08351813.2016.1196544.

Wells, G. (1993). 'Reevaluating the IRF sequence: A proposal for the articulation of theories of activity and discourse for the analysis of teaching and learning in the classroom'. *Linguistics and Education*, 5(1), 1–37. https://doi.org/10.1016/S0898-5898(05)80001-4.

Wetherell, M. (2007). 'A step too far: Discursive psychology, linguistic ethnography and questions of identity'. *Journal of Sociolinguistics*, 11(5), 661–81.

Wiliam, D. (2011). *Embedded Formative Assessment*. Bloomington, IN: Solution Tree.

Wong, J., & Waring, H. Z. (2009). '"Very good" as a teacher response'. *ELT Journal*, 63(3), 195–203. https://doi.org/10.1093/elt/ccn042.

Wood, D., Bruner, J. S., & Ross, G. (1976). 'The role of tutoring in problem solving'. *Journal of Child Psychology and Psychiatry*, 17(2), 89–100. https://doi.org/10.1111/j.1469-7610.1976.tb00381.x.

Wood, T. (1994). 'Patterns of interaction and the culture of mathematics classrooms'. In S. Lerman (ed.), *Cultural Perspectives on the Mathematics Classroom* (pp. 149–68). https://doi.org/10.1007/s13398-014-0173-7.2.

Wood, T. (1998). 'Alternative patterns of communication in mathematics classes: Funneling or focusing?' In H. Steinbring, M. G. Bartolini Bussi, & A. Sierpinska (eds), *Language and Communication in the Mathematics Classroom* (pp. 167–78). Reston, VA: National Council of Teachers of Mathematics.

Wood, T., Cobb, P., & Yackel, E. (1993). 'The nature of whole-class discussion'. *Journal for Research in Mathematics Education*, 6, 55–68.

Wooffitt, R. (2005). 'Conversation analysis and discourse analysis: A comparative and critical introduction'. *The British Journal of Sociology*, 57(4), 234. https://doi.org/10.1111/j.1468-4446.2006.00133_21.x.

Wowk, M. T. (2007). 'Kitzinger's feminist Conversation Analysis: Critical observations'. *Human Studies*, 30, 131–55. https://doi.org/10.1007/s10746-007-9051-z.

Xu, L., & Clarke, D. (2013). 'Meta-rules of discursive practice in mathematics classrooms from Seoul, Shanghai and Tokyo'. *ZDM: Mathematics Education*, 45(1), 61–72. https://doi.org/10.1007/s11858-012-0442-x.

Yackel, E., & Cobb, P. (1996). 'Sociomathematical norms, argumentation, and autonomy in mathematics'. *Journal for Research in Mathematics Education*, 27(4), 458–77.

Yackel, E., & Rasmussen, C. (2002). 'Beliefs and norms in the mathematics classroom'. In G. C. Leder, E. Pehkonen, & G. Törner (eds), *Beliefs: A Hidden Variable in Mathematics Education?* (pp. 313–30). https://doi.org/10.1007/0-306-47958-3_18.

Yerrick, R. K., & Roth, W.-M. (eds). (2005). *Establishing Scientific Classroom Discourse Communities: Multiple Voices of Teaching and Learning Research.* https://doi.org/10.4324/9781410611734.

Zimmerman, D. H. (1998). 'Discoursal identities and social identities'. In C. Antaki & S. Widdicombe (eds), *Identities in Talk* (pp. 87–106). London, UK: SAGE Publications Ltd.

Index